To Understand a Cat

Advances in Consciousness Research (AiCR)

Provides a forum for scholars from different scientific disciplines and fields of knowledge who study consciousness in its multifaceted aspects. Thus the Series includes (but is not limited to) the various areas of cognitive science, including cognitive psychology, brain science, philosophy and linguistics. The orientation of the series is toward developing new interdisciplinary and integrative approaches for the investigation, description and theory of consciousness, as well as the practical consequences of this research for the individual in society.
From 1999 the Series consists of two subseries that cover the most important types of contributions to consciousness studies:
Series A: Theory and Method. Contributions to the development of theory and method in the study of consciousness.

Volume 70

To Understand a Cat: Methodology and Philosophy
Sam S. Rakover

To Understand a Cat

Methodology and philosophy

Sam S. Rakover
Haifa University

John Benjamins Publishing Company
Amsterdam / Philadelphia

™ The paper used in this publication meets the minimum requirements of American National Standard for Information Sciences – Permanence of Paper for Printed Library Materials, ANSI Z39.48-1984.

Library of Congress Cataloging-in-Publication Data

Rakover, Sam S., 1938-
To understand a cat : methodology and philosophy / Sam S. Rakover.
p. cm. -- (Advances in Consciousness Research, ISSN 1381-589X ; v. 70)
Includes bibliographical references and index.
1. Cats--Behavior. 2. Cats--Psychology. 3. Philosophy of mind. 4. Science--Methodology. I. Title.
SF446.5.R35 2007
153--dc22 2007013284
ISBN 978 90 272 5206 7 (Hb; alk. paper)

John Benjamins Publishing Co. · P.O. Box 36224 · 1020 ME Amsterdam · The Netherlands
John Benjamins North America · P.O. Box 27519 · Philadelphia PA 19118-0519 · USA

Table of contents

Preface

About two and a half years ago, when my granddaughter Tai was about to graduate first grade, I decided that it was time I wrote a children's book for her. What should it be about? What was likely to spark Tai's interest? I mulled several possibilities over in my mind, and then in a flash of inspiration I decided I would write about the entertaining doings of Max, a beautiful Himalayan cat, given to me as a gift by my daughter Shelli, Tai's mother. At once I began reviving a fascinating episode of Max (see the beginning of chapter one). But even as I wrote it occurred to me that this episode of Max was not quite right for children of Tai's age, and that I was actually in up to the neck in the interesting-bugging question of how to explain Max's behavior, and that my thoughts for years on psychology as a science and on the mind/body problem were engulfing me like a tidal wave: thus I found myself writing the first draft of this book. (In fact, thoughts about the mind/body question have assailed me since that day when Professor Leibowitz, my master and teacher, argued in the first year of my studies in the Psychology department at the Hebrew University in Jerusalem that contrary to phenomena treated by the natural sciences, the phenomenon of his toothache, Professor Leibowitz's own, was subjective, and no one other than he himself could feel it.) Afterwards I was preoccupied with other matters and when I got back to the book I rewrote it from start to finish. So what is the book about?

The book tries to find solutions to the three following problems.

First problem: The need for mentalistic explanations. In a fairly short time it became clear to me that to explain a large group of behavioral episodes of Max the cat, I had to use a special kind of explanations, which I call "mentalistic explanations", that are based on mentalistic concepts such as will, purpose, intent, and consciousness. This was because it was impossible to understand his behavior, so it seemed to me, by use of mechanistic explanations prevalent in the natural sciences, which would treat him as a machine or as a creature constructed of a collection of reflexes and instincts only. Here it is worth clarifying briefly what a scientific explanation is.

Scientific explanation: I cannot review this complex subject as part of the preface (see later on), so I shall describe very briefly my outlook on the matter. I maintain that an incomprehensible phenomenon obtains a scientific explanation, and becomes understood, when we mesh, integrate (in certain forms that I shall not discuss here) the given phenomenon into the framework of a system that we already understand, a system firmly anchored to broad theoretical and empirical ground. That is, I suggest that science explains a not-understood thing by means of an understood system, because something not understood cannot be understood by means of something that is also not understood. I shall call the understood system, by which the not understood is explained, an "explaining system", and the not understood phenomenon the "phenomenon to be explained". To illuminate this matter I shall give two examples. In the morning the car won't start. The mechanic (who has studied electricity and its applications in a car) lifts the hood of the engine and finds that the battery is dead. He replaces the old battery with a new one and the engine starts easily. The explanation for the non-movement of car, for the phenomenon to be explained, is accomplished by meshing, integrating this phenomenon into the car's electrical system – the explaining system. Ronen is coughing and he has a temperature. The doctor examines him and says that Ronen has pneumonia; he treats him with antibiotics and Ronen gets well. In this case too the explaining system (which is understood by doctors, of course) suggests an explanation for the not understood phenomenon (a cough and a temperature). That is, the phenomenon has an explanation. These are relatively simple explanations that are widespread in the natural sciences: mechanistic explanations.

However, in addition to the mechanistic explanation there is a different kind of explanations, which we use in everyday life. For example, we explain David's trip to Tel Aviv by noting that this act of David's (traveling to Tel Aviv) realizes his will to see Verdi's opera *Falstaff*, which is to be performed at the Tel Aviv Opera House. This explanation is based on David's subjective will/belief, so the explanation differs in this respect (and in others, which this book discusses) from the mechanistic explanation. As stated, I call this a mentalistic explanation. What is important to stress here is (a) the phenomenon to be explained in the present case is David's action: this is the phenomenon that we are required to explain, namely why David traveled to Tel Aviv; (b) The explaining system appeals to the will to see *Falstaff* and to the belief that a journey to Tel Aviv will realize this will. This will/belief therefore is an important part of the explaining system. Here, however, the following question arises: why is this explaining system understood?

The answer is because regular and normal everyday life for the individual, her will, and her belief (where this will be realized by means of a certain action) are mental states that are clear and self-evident. Why did you register for the Psychology Department? Because I want to be a clinical psychologist and I believe that the way to realize this goal is by registering for the Psychology Department. The explanation for a behavior in this case, and in other similar cases, is brought about by an appeal to the individual's mental world, the world that to the individual herself is understood and clear – these are her intentions, and these, truly, are an important part of her very being.

Here one is likely to wonder, is that really so? May it not be asked what the causes of will and belief are? Psychology of the personality and psychoanalysis show by a very interesting analysis that not always does the individual understand himself, and in certain cases it transpires that the individual's will/belief are nothing but an expression of deep and unconscious desires. Is this not the case? My answer is: true, in these cases it is so. What is thought to be understood proves in the end to be not understood. But here I go on to argue that to explain this new not understood phenomenon we have no choice but to appeal to a new explaining system, a system that must be understood (because something that is not understood cannot be explained by something else that itself is not understood), by means of which we shall be able to explain the new not-understood phenomenon. (Sure enough, Freud built the theory of psychoanalysis as an explaining and understood system, a system whereby it would be possible to understand mental behaviors that only seemed to us to be understood until Freud revealed that in fact they were not understood and needed a new explanation.)

This process by which the seemingly understood turns out to be not understood is a research process characteristic of empirical science as a whole. It emerges that the explaining systems in the natural sciences also undergo changes in understanding; and what once was thought to be understood proves following further research to be a complex phenomenon requiring a new explanation by an appeal to a new understood-explaining system. This apparently is an interminable process, constituting an important part of the development of science. For example, Galileo's and Kepler's laws received an explanation in the framework of Newton's theory of gravity, and the concept of gravity received an explanation and a new scientific meaning in the framework of Einstein's theory.

Now let us return to our main concern: the need for mentalistic explanations to understand the behavior of Max the cat. It seems to me that Fodor (1987) may have felt something similar when he tried to explain the behavior of Greycat, his cat, by means of mentalistic concepts; he writes:

> The theory is that Greycat – unlike rocks, worms, nebulas, and the rest – has, and acts out of, beliefs and desires. The reason, for example, that Greycat patrols his food bowl in the morning is that he wants food and believes – has come to believe on the basis of earlier feedings – that his food bowl is the place to find it. (p. x)

The problem is that in the view of many researchers, explanations of this kind, mentalistic explanations, are not perceived as scientific: they are deemed anthropomorphist, that is, explanations that unjustifiably humanize animals. As a result, I began to explore whether this kind of explanations could go under the umbrella of the accepted methodological rules of the game in the natural and social sciences, namely be made to satisfy the requirements of empirically scientific explanation and research. Therefore, I give the name "Scientification" to the approach that modifies the concepts of mentalistic explanations such that they fulfill the methodological requirements accepted in the natural and social sciences.

As will become clear later, scientification is methodological in essence and different from other scientific approaches to goal-directed explanations, that is, explanations based on belief/desire. For example, those approaches perceive these explanations as a kind of theory of folk psychology or commonsense belief/desire psychology (Fodor, 1987, p. x), justified and defended by scientific psychology (cognitive psychology and neurophysiology); they base folk theory on cognitive psychology and neurophysiology, and also see folk psychology as a particular kind of empathic process by means of which it is possible to understand the self and the other. Fodor's (1994) major philosophical project is to reconcile the idea that psychological explanations are intentional with the idea that mental processes are computational. The main problem that this project faces is solving the following crux. Given that the mind is a computer of some sort or other, how is it possible to generate the mind's meaningful and broad content that represents the world by employing mechanistic computational processes, symbol-to-symbol transformations, syntax properties? Resolving this, however, is not the present book's project.

In contrast to Fodor's project, this book's project is based on the realization that currently there seems to be no solution to the problem of the relation between the intentional and the computational (or between the intentional and the neurophysiological). Given this, and the understanding that an adequate explanation of behavior involves the employment of mentalistic concepts (common in folk psychology) and mechanistic concepts (common in the natural sciences), this book contends with the following question. How may an approach be constructed that will employ these two different and irreconcilable concepts in a coherent explanatory manner? How may both mentalistic explanations (prevalent in folk psychology) and mechanistic explanations (prevalent in the natural sciences) be employed in a unified theoretical approach that fulfills the methodological requirements of science? The book's aim is to find a solution to these questions, and not to suggest a new answer to the problem of the intentional and the computational (or the intentional and the neurophysiological). As mentioned above, one central pillar of the present approach, scientification, is to find a way to modify mentalistic explanations such that these will satisfy the requirements of scientific methodology. That is, scientification suggests perceiving mentalistic explanations as specific explanations created by means of explanatory mentalistic schemes, or special explanatory models, which meet the accepted methodological requirements in the natural and social sciences. The concept of scheme or model of scientific explanation requires a brief clarification.

A scheme (model) of a scientific explanation: A scheme of explanation is a general procedure by means of which specific explanations are created for specific not understood phenomena. Not every explanation, and not every scheme of explanation, is right from a scientific viewpoint. The scientific explanation has a special structure, which must satisfy two fundamental requirements: rationality and empiricism. One of the most important realizations of these two requirements is the proposal that the scheme of the scientific explanation, by means of which specific explanations are given to empirical observations, must have the structure of a logical argument (see Hempel, 1965). That is, science stipulates that the scheme by means of which scientists create specific empirical scientific explanations for different observations must possess the structure of a logical argument. The scheme of the scientific explanation has to include the derivation of the description of the not understood empirical phenomenon from a scientific theory or scientific law, just as we logically draw a conclusion from clear and understood assumptions. In other words, one of the important schemes or models for a scientific explanations proposes that the description of the not understood empirical phenomenon will obtain an explanation when we are able to derive this description from a theory (i.e., from the explaining system) deductively. (As stated, this scheme of explanation is only one of the schemes, models, acceptable for a scientific explanation. On the problems of this scheme, and on other alternative explanation schemes that appear in science, see below and in Rakover, 1990.)

Second problem: Matching of the mechanistic and mentalistic explanation to behavior. I realized that Max the cat's behavior is extremely complicated. It is a behavioral system constructed of both behavioral components that require mentalistic explanations, that is, explanations that address the cat's inner conscious world: desire, knowledge, purpose, intention; and of components that require mechanistic explanations, that is, explanations acceptable in the natural sciences and in cognitive science, namely neurophysiological explanations and explanations based on the computer analogy. Given complex behavior, how may we know what to explain it by? By turning to the mechanistic or the mentalistic explanation? Or to both? Is it possible to develop criteria that will suggest how a behavior and the kind of explanation can be matched?

Third problem: Mind/body. The two foregoing problems could be resolved were it possible completely to reduce the mentalistic explanation to the mechanistic, or to suggest a perfect explanation for the mind in terms of the body (neurophysiology of the brain). In this case scientists would use only mechanistic explanations. As in the natural sciences, where different kinds of mechanistic explanations are employed for different phenomena, different kinds of mechanistic explanations will be employed in psychology. But it became clear to me from extensive reading of the appropriate literature, and devoting a great amount of thought to the question, that at present no satisfactory solution appears on the horizon to the eternal question of the unknown connection between brain and mind. In the present state of knowledge, then, I maintain that there is room for an approach based on mechanistic and mentalistic explanations together, because this approach will supply greater understanding than that provided by each of them separately.

After contending with these three problems, I decided to develop a relatively new methodological approach, which I call "methodological dualism", whose principles are the following.

a) The methodological status of mentalistic and of mechanistic hypotheses is equal from the viewpoint of an empirical test. Still, and for other methodological reasons, it is recommended to use hypotheses suggesting mentalistic explanations when they raise our understanding of the studied behavior above what is suggested to us by mechanistic hypotheses. That is, it is recommended to use a mentalistic hypothesis when it provides an explanation for that part of the behavior that is not accounted for appropriately by a mechanistic hypothesis. Mentalistic hypotheses are based on the assumption that the individual (human or animal) is imbued with mental states and processes (desire, belief, goals, intentions, thoughts, emotions, and so on).

b) It is possible to use specific mentalistic explanations created by mentalistic explanatory schemes (procedures for giving specific explanations). These schemes meet an important part of the requirements of scientific methodology for proposing explanations that prevails in the natural and social sciences. Science explains various phenomena by use of explanatory schemes that possess certain methodological properties; mentalistic schemes possess these methodological properties, so they may be seen as belonging to scientific methodology, even though they cannot be reduced to mechanistic explanatory schemes. The present approach, therefore, does not treat day-to-day explanations, which resort to desire, knowledge, intention, as folk theory reminiscent of scientific theory, or as an empathic, psychological process, in which one conducts mental simulation of the other, but as mentalistic explanatory schemes, procedures that yield specific explanations for a specific mental behavior.
c) Complex behavior, which in psychology is usually explained mechanistically, is by the present approach explained through the "multi-explanation theory", which unites in a coherent manner the use of mentalistic and mechanistic explanatory schemes. This coherence is attained by matching the kind of explanation with the kind of behavior. In the natural sciences an explanatory model or scheme utilizes laws or various theories to propose an explanation suited to different phenomena; by contrast, the multi-explanation theory proposes that a theory that attempts to understand complex behavior must use several explanatory models – mechanistic and mentalistic. Even though the explanation offered by the multi-explanation theory is imperfect, because it cannot and does not aspire to offer a solution to the mind/body problem, the understanding that this theory offers is greater than that provided by the mechanistic or the mentalistic explanation separately.

Methodological dualism, then, is not an approach that offers a metaphysical-ontological solution to the mind/body problem, but a methodological approach that proposes methods to construct theories of behavior that use mechanistic and mentalistic explanatory schemes together. Multi-explanation theories are not just a kind of instrumentalist theories, namely efficient calculating machines for predicting behavior, because the concepts of these theories represent mentalistic, cognitive, and neurophysiological states. Despite the thorny problems associated with mentalistic concepts, I propose that concepts that represent conscious processes, such as desire/belief, enrich and strengthen the multi-explanation theory in both its explanatory power and its openness to empirical tests (see discussion on these matters later, particularly chapter 9).

This dualism was indeed applied to the Max the cat's behavior as an individual case and developed by an analysis of his behavior, but the methodology was also applied to the behavior of other animals, including humans. In other words, most of the behavioral examples treated in the book are in fact associated with Max, but there is also reference to human behavior and behavior of other animals. As a result of these applications, and a comparison of the methodological dualism with relevant scientific and philosophical approaches, the above three principles were developed, and they were fleshed out in the form of the nine chapters of the book.

Chapter 1: This chapter proposes dealing with mentalistic and mechanistic explanations as alternative explanatory hypotheses for a given behavior, where the methodological status of these hypotheses is equal from the viewpoint of an empirical test. Furthermore, the chapter alludes to additional elaborations, which follow in the rest of the book, connected to such problems as the contribution of the mentalistic explanation, the decomposition of behavior into its components, and the match between the kind of explanation and a given behavior.

Chapter 2: This chapter recommends a methodology for constructing hypotheses out of behavioral episodes, and putting these hypotheses to empirical tests. The basic idea is to propose as a hypothesis the best explanation for a given behavioral episode, and to generalize it over such variables as animals, situations, and responses. The chapter also suggests a criterion, the "new-use principle", whereby the mentalistic explanation is to be applied to a given behavior when the latter is characterized by the attainment of the same goal through various responses or of different goals through the same response.

Chapter 3: This chapter suggests justifications for explaining the cat's behavior by mentalistic explanations, namely by appeal to his mentalistic-conscious world. To support the attribution of consciousness to the cat, a behavioral criterion for free will (connected to the new-use principle) was developed, which may be applied to the cat's behavioral episodes. Now since the behavior of free will is conscious, it is inferred by analogy that the cat's behavior likewise is conscious. In addition, the chapter discusses evidence and arguments supporting the idea that animals are imbued with different degrees of consciousness.

Chapter 4: This chapter offers an answer to the question of why the methodological status of the mentalistic explanation is lower than that of the mechanistic explanation, even though all the hypotheses are equivalent from the viewpoint of an empirical test. The answer lies in a comparison of the structure of the mechanistic and the mentalistic theories, in which flaws were discovered regarding the inner consistency of the mentalistic theory and the connection between its theoretical concepts and observations. Despite these flaws, the mentalistic explanation has enormous importance, because it was not possible to suggest an exclusive mechanistic explanation for a wide group of daily behaviors of Max the cat. This group meets the criterion of mentalistic behavior, but does not meet the requirements of the mechanistic criterion, which is met by another group of behaviors that are explained by computational, neurophysiological, innate, and evolutionary processes.

Chapter 5: This chapter proposes an explanatory system, the "three-stage explanation", for explaining complex behavior of a cat, constructed out of mentalistic and mechanistic components. The underlying idea is to break a given behavior down into its behavioral components, and to focus on one central behavioral component, which constitutes the axis of the explanation. At the first stage a goal-directed mentalistic explanation is adapted to this behavioral element; at the second stage this element is explained in accordance with its mechanistic function in the behavior that preceded the behavior under study. This function is standard and typical of the cat's behavior (e.g., explaining the cat's claw-sharpening according to the innate survival function); at the third stage it is explained how this component is integrated into the goal-directed mentalistic explanation (which is set out in first stage) by the acquisition of a new role, a new purpose, namely by highlighting the functional change that this component has undergone, from a mechanistic function in the previous behavior to a new role in the investigated behavior – a role that serves the mentalistic explanation.

Chapter 6: This chapter proposes a "multi-explanation theory", intended to address complex behaviors that can be broken down into a network of behavioral components that require mentalistic and mechanistic explanations. In the natural sciences the model of explanation uses several laws or theories to suggest an explanation for a given phenomenon, while in psychology the relationship is the reverse: in psychology the theory uses several mechanistic and mentalistic explanatory models. The multi-explanation theory may create three problems that might undermine its operation: it gives an ad hoc explanation, it lacks inner consistency, and it cannot be compared with different theories. The chapter offers procedural guidelines that resolve these three problems. Still, it transpires that while the multi-explanation theory can be put to an empirical test like any theory in the sciences, there is an important difference between the way a theory in natural sciences proposes an explanation and the way in which the present theory functions. This difference lies in the fact that the multi-explanation theory is based on two kinds of explanatory models: mechanistic and mentalistic.

Chapter 7: This chapter proposes a methodological foundation for the multi-explanation theory, by the scientification approach. The chapter shows that specific goal-directed explanations (desire/belief) are produced from a goal-directed mentalistic explanatory scheme, which operates according to the accepted rules of the game in science. Just as specific explanations in science are based on mechanistic explanatory schemes (e.g., Hempel's D-N model), so specific explanations of behavior of human beings and animals are partly based on a goal-directed mentalistic explanatory scheme. The chapter shows that this explanatory scheme upholds several methodological properties of explanatory schemes in the sciences, for example, the property that the refutation of a specific mentalistic explanation has no implications for the explanatory scheme itself. Moreover, the chapter reveals that the goal-directed explanatory scheme (desire/belief) is not a kind of scientific law, because this scheme does not hold the features of a law acceptable in the sciences.

Chapter 8: This chapter posits a theoretical basis for the multi-explanation theory by presentation of the reasons why we still have not succeeded in proposing a solution to the mind/body problem. If it were possible to reduce awareness, consciousness, to the neurophysiology of the brain, there would be no justification for the multi-explanation theory, because all would be explained in terms of the natural sciences. The chapter examines five main research areas connected to the mind/body problem: mental causality, functionalism and the multiple realizability argument, the computer and the process of breakdown into basic mechanisms, neuropsychological reduction, and consciousness; it concludes that indeed a scientific approach or philosophy has yet to be found that offers an acceptable solution. The chapter also suggests several theoretical and empirical reasons why consciousness is not an epiphenomenon but an important and essential factor in the explanation of behavior.

Chapter 9: This chapter tries to fathom the uniqueness of the approach that I developed here by comparing it with relevant approaches in the philosophy of science and mind. First, it compares methodological dualism (based on the scientification approach: granting a methodological stamp of approval to mentalistic explanations) with explanatory dualism, functionalism, and levels of explanation. Second, it compares the multi-explanation theory with other approaches to the construction of a scientific theory. Third, it discusses the question: what kind of understanding does this theory supply? Finally, the chapter discusses whether consciousness can be viewed as an "emergent property".

As can be seen from the chapter summaries, the book is organized in the following way. In the first three chapters the behavior of Max the cat is handled methodologically. In chapters 4 and 5 descriptions of the cat's behavior diminish, while the discussion of the philosophy of science (explanation and theory) and of mind broadens. From chapter 5 to 9, the development of methodological dualism and multi-explanation theory takes center-stage. Accordingly, the methodological-philosophical discussion expands and deepens as the reader progresses through the book.

And after all that work was done, doubts arose and welled up in me until they opened within me a dark window to an understanding of why Gogol burned the second part of "Dead Souls". The doubts had been there all the time, but I had suppressed them so that they would not stop up the flow of ideas and the writing; and I had opened a small aperture for them from time to time to use them to improve and shine up the ideas. But now, the job done, the doubts burst their banks, black and muddy water cascading over the dam, carrying the threat that I might commit the same act as the great humorist writer. And then I decided to outsmart them and I wrote them in the form of aphorisms, one at the beginning of each chapter.

Heartfelt thanks go to a large number of people who read the book, made very important comments, and helped in its improvement: Maxim Stamenov (the editor of this series), Dany Algom, Adir Cohen, Amotz Dafny, Morris Goldsmith, Itzhak Hadani, Meir Hemmo, Ido Landau, Shimon Marom, Israel Nachson, Aviva Rakover, Ruth Ramot, Saul Smilansky, and an anonymous reviewer.

Special thanks go to Murray Rosovsky who translated the whole book from the original Hebrew.

I wish especially to thank architect Avner Oren, who made the drawing of Max's living space (the Rakover family apartment).

CHAPTER 1

Scientification

Placing anecdotes and anthropomorphism under the umbrella of science as the first step

The purpose of this chapter is to suggest a methodology that will treat anecdotes and anthropomorphist explanations in a way acceptable to science. I call this proposal "equal hypotheses testing". By this approach, mentalistic hypotheses, which ascribe to an animal mental states and processes similar to our own, have to be compared empirically with mechanistic hypotheses, which explain the animal's behavior as if it is a neurophysiological machine. The methodological status of each of these kinds of hypotheses when put to the scientific empirical test is the same. The chapter starts with an anecdote describing Max the cat preparing an "ambush for the night moth".

I fully believe that Max the cat has a soul, consciousness, and all that, but in the middle of the night when I am awoken by his yowling, I look angrily for the switch to turn him off.

1.1 An ambush for a night moth

Max is a Himalayan breed of cat, a hybrid of a Siamese and a Persian. A very handsome male, loved and spoilt. His eyes are a shining blue, like a summer's pure sky. His nose, ears, paws, and splendid tail are black, and the fur of the rest of his body is whitish golden with grey patches. His cheeks are adorned with impressive black fur, which imparts to the picture of his head the shape of a great trapeze, above which rise two black triangles, his graceful ears. His fine face, then, is heart-shaped, and this, I am persuaded, is because Max is truly a good soul.

One evening, in midsummer, I am lying on my side in front of the television, gazing at the flickering screen. My head rests on my left palm, as I doze off from time to time. The August night is very hot. The night moths flutter around the porch lamp hanging over the large guest table. Occasionally they strike the paper lampshade, with the sound of an impact, a collision, pit, pat, pit-pit, pat-pat, pat-pit, and they continue their interminable circling in the light radiated by the bulb. I turn my head toward them and ponder the meaning of this dance spinning around the lamp. There are about seven moths swooping and swirling round and round to the rhythm of the pit-pat of the collisions. I have no idea why night moths love to dance like that around the porch lamp. My look strays into the dark of the night beyond the porch and turns into a stare.

What would a moth think if it were to come across a dance floor filled with teenagers? Let's assume for a moment that it landed on the electric flex of the strobe light and looked at the dancers below. The music thunders out at a beat, brum-brum-brum, bom-bom-bom, the lights scintillate in a welter of colors and those strange beings below constantly make odd noises, yelling and screaming, spinning in circles and moving in weird twists as if struck by a terrible storm. Does the moth distinguish males from females? Does its tiny head between its gorgeous wings grasp that down below a strange and complex ritual of courting is going on, the boys after the girls and the girls after the boys? Perhaps, I think to myself, the male moths are wooing the female moths in a circle dance around the porch lamp? Maybe my porch has turned into a dance hall for young moths? I am unable to distinguish a he-moth from a she-moth.

Suddenly in the corner of my eye I discern interesting behavior by Max the cat, which I shall call **the ambush for the night moth** (important note: from now on I shall signify Max's behaviors in **boldface letters**). Max looks hard at the swirling moths, flies, birds, cars, and people hurrying in the street, with prolonged concentration. Sitting on his rump, his sumptuous tail curled around him, his two forelegs tense, his head turned up to the lamp and his ears pricked. His black pupils are large in the night, and each one occupies his beautiful eyes almost entirely. The posture of a regal seated sphinx, motionless, as if carved in marble, only the tip of his tail slowly moving right and left. Max is thinking. I am convinced that he engrossed entirely in thought. This can be seen in the tension along his whole body and at the end of his moving tail. I can actually see the wheels of his thinking turning in absolute silence, there in his fine head. But what is he thinking about? He is following the moths' dance, of this I have no doubt. But is he thinking, like me, that they are dancing? Just reeling around in the lamplight? Or is he following the circular motion of flying objects? What is going round in his head? Maybe he is simply staring at the movement of the moths, and maybe he is mesmerized by the shadows of the dancers cast on the wall and the ceiling.

When I awake from my doze, the cavorting figures on the TV screen are still kicking each other in the face, firing guns with a horrifying racket, rat-ta-ta-ta-ta-tat, and blowing up cars and buildings. Nothing new. All the moths have retired from the dance and fluttered off the porch into the darkness. Max, now ensconced on a chair in front of the TV, shoves his nose into his tail and drifts off like me. Suddenly a noise is heard of a collision inside the lampshade on the porch, pit, pat, pit-pit, pat-pat, pat-pit. Simultaneously we both turn our heads to the porch, and there, as we expected, we see one big moth doing the spinning dance around in the lamp. Suddenly it dives, and circles low around the dining table on the porch. Max leaps from the chair, chases the flying insect, shoots his forelegs out at the moth, which dodges every feline slice at the very last instant as Max punches out his black paws in a succession of left and right jabs, claws unsheathed. Claws curved like an eagle's beak, tempered and sharp. His body movement during the chase after the moth is lithe, light, smooth, powerful. How lovely and pleasing is the dance of the predator Max. Predator? Could he perhaps just be enjoying himself, playing? Could he not at one fell swoop kill the moth, whose flight, for all its grace, suddenly seems to me slow, turgid, and cumbersome, compared with Max's stunning agility, with which I am painfully acquainted from my own games with him? The countless scratches on my right arm, my strong, quick arm, will attest like a thousand witnesses to the speed of the blows that Max can land when he wants. No doubt about it, Max is pulling his punches, and he hasn't the slightest intention of wasting the moth. Max is having fun.

Suddenly the moth sweeps upward, and stays hanging upside-down off the porch ceiling. Max stops, his head turning rapidly right and left, up and down, forward and back (yes indeed, Max's flexibility is way beyond what we are capable of), but the night moth has slipped out of his field of vision.

And now, I have to say, Max begins to perform a series of amazing actions. He starts to sniff all around, stretching his supple body farther and farther. And again he turns his head back and forth, combing the area. Again he sniffs, his head flat down between his shoulders, his nose barely a fraction off the floor; he rounds the table legs and returns to the exact spot where he was when the moth disappeared. He remains in his noble sphinx-like posture, raising his head toward the porch light. And then he does something astonishing! He jumps onto the dining table, sits in his sphinx-like posture beneath the porch light, raises his head to the lamp once again, squinting at it with a concentrated and steady gaze. Thus he sits silent for a while, waiting with sublime patience, his head turned up to the light. The end of his tail twitches slowly to the left and to the right, while I hold my breath, amazed, watching Max as he waylays the nocturnal moth.

You've got to admit, I say to myself later, as I hoist the cat up, taking a long look at his fine furry face, now totally bland, that his eyes and visage show no sign that he is capable of such a complex thinking act. Everyone tells me that cats are stupid and selfish, and live only by their instincts. And here is my Himalayan cat doing the unbelievable: he draws a logical conclusion from earlier information, from earlier understanding, a conclusion that leads him to set a trap for the night moth, which has suddenly disappeared in the middle of his fun chase. Think for a moment about what Max had to understand and to know in order to lay his ambush.

First, Max realized that night moths love to dance around the porch light. Next, Max realized that the moth, which before had danced around the light – pit-pit-pat-pat, and which had got away from him in the middle of the enjoyable chase and left him frustrated, belongs to a group of moths that like to fly around a light. And now, given these two realizations, Max inferred that the runaway moth was likely to resume its flight around the porch light bulb, and that there, on the table under it, and within leaping distance, he could grab it and continue his entertaining chase.

I am gripped in amazement; I kiss Max on the head between his two black ears, ease myself into the armchair in front of the TV, and set the cat on my knees. Max stretches and at once begins to purr in pleasure. He deserves it, I'm thinking. My right hand strokes his soft fur and meanders up and down, I gaze unseeing at the flickering

TV screen and wonder whether Max thinks in the language I think in. Because the thoughts that I ascribe to the cat I express in Hebrew. Because I speak my thoughts by means of words and sentences in Hebrew. Max certainly cannot formulate his thoughts in Hebrew like me, a human being. So how does he express them? What is his cat language? Does that language make him able to reach conclusions similar to those that I have reached, conclusions that I attribute to him as if he, Max, has stated them? I have no doubt that Max in his actions has expressed highly intelligent behavior, which even very young children are incapable of attaining. "Max," I say to him, cupping his head between my two palms, peering into his jet black pupils, "tell me your secret, Max. What goes on in this handsome head of yours?"

The life of Max, I continue, reveling in my thoughts, is based principally on the sense of smell. Our life, human life, depends primarily on vision and hearing. We express out thoughts by means of these two senses. We talk to each other and read the written text. These are the vehicles of our thought. Because Max depends on the sense of smell, are his thoughts ordered by means of some olfactory code? Imagine that Max draws conclusions based on complex relations between different and fine smells. The smell of garlic together with the smell of onion tells Max that it is time to snooze. And the smell of evening rising from the soil, along with the smell of the lamp given off by its heat, tells him that it is time to chase moths. It seems fantastic to me. How on earth can Max be understood?

These thoughts get me nowhere, and I glance from time to time at the television which continues to show death penalties being carried out. Max gets off my knees and jumps onto the sofa, lies on his side and rests his head on the cushion (**head on the cushion**). Amazing. This is the first cat I have seen that sleeps with a pillow. I have no doubt that he learned it from me or from Aviva. Is it because Max lives by the sense of smell that he cannot enjoy a beautiful picture or good music? I have noticed that Max goes on sleeping in perfect tranquility when we, my wife Aviva and I, are listening to *Don Giovanni*, to Mahler, or to Berlioz, but he pricks up his ears when he hears the howl of a cat coming from the TV (**musical indifference**). I have also noticed that Max looks at his reflection in the big mirror standing in our guest room. At first he sniffed at the image in the mirror, and checked out what was behind the mirror too, and when he did not smell a cat in the mirror, or find a cat behind it, he lost interest – although occasionally he would go and take another look at the mirror (**cat in the mirror**). I believe that enjoyment of beauty is linked to vision and hearing, but not to smell. Different combinations of sounds or of colors arouse different feelings of pleasure in me. This is very beautiful music, I say, and this music is not beautiful, it jars on the ear. Please turn off the radio. Max has a handsome face, and that cat has an ugly face. But never in my life have I come across a combination of different smells that arouses in me a sense of beauty. A combination of different smells creates nothing but a new smell, pleasant or not, but in no way does this combination of different smells engender in me a feeling of beauty. It would be quite odd to say that a certain smell is beautiful, as we say of the Mona Lisa.

Again, thoughts galloping into nowhere. Back to Max. What, in fact, is so astonishing about **the ambush for the night moth**? Am I excited because Max the cat, a creature usually considered made of bundles of instincts, has performed a behavior that exceeds the expected? What am I getting so worked up about? I ask myself, and I answer. If, for example, you, the reader were watching one of those British detective movies, say Sherlock Holmes, Hercules Poirot, or Miss Marple, would you not marvel at how smart this elderly lady sleuth is, when she, by virtue of prior information based on sundry evidence, sets a trap for the crook and catches him red-handed? Would you not be amazed? You would certainly marvel at her savvy and ingenuity. I know I would.

It's not the same thing, you may say. With Max this is just instinct, innate behavior, without the slightest sign of intelligent thought. Instinct! I look at Max sleeping on the sofa: is that all? Were you born that way? Did nature form you thus, Maxie? The hunting instinct? To leap onto the dining table on our porch to hunt a night moth that you chased before without success? Has evolution worked so hard for millions of years only so that this Max can prey on the poor night moth? A perfected hunting machine, whose entire essence boils down to chasing night moths pit-patting on the porch lampshade? I don't think so, I said to myself, and changed the TV channel.

1.2 Some methodological thoughts: Anthropomorphism and anecdotes

A professional psychologist, trained like me in the conduct of laboratory experiments (learning of fear and avoidance in rats, perception and recognition of faces), may turn her nose up at reading the foregoing passage. Listen, she will say, this isn't a scientific description and it isn't a scientific explanation. It's an anthropomorphic anecdote in the best case, and literature in the worst. Why? Let's take a look. This criticism has two aspects. First, anthropomorphism is not a *proper scientific explanation* because the explanation for Max's behavior by attribution of human mental processes, as we shall see, is not justified and stands on rickety foundations. Second, the anecdote of **the ambush of the night moth** is not a *scientific observation* because this description does not satisfy the accepted scientific requirements.

1.2.1 Scientific observation and anecdotes

Psychology has undertaken to play the science game according to several methodological rules acceptable to the natural sciences. Some of these rules are linked to the *performance of observations* and some to *giving explanations* for these observations. An *observation* is deemed scientifically sound when it meets in principle the following three requirements: publicity (public availability), objectivity, and repeatability. Publicity requires that all scientists can observe the given phenomenon, so that the description of it will not depend on a particular observer, her economic or political authority, or her leanings. Objectivity requires that the observation will not be influenced by the observer's attitude, and the reverse – that the phenomenon will not influence the observer, so that phenomenon will be described as it is, that the description will be valid, and that science will not supply explanations for biased or distorted phenomena. (Application of this requirement to quantum observations raises several problems that exceed the scope of this book.) Repeatability requires that it is possible to repeat observing the phenomenon several times, so that we may be able to examine it from every possible angle and ascertain that there is no pure chance here; and so that it will be possible to conduct experiments to test different explanations of the phenomenon.

Does the above description of **the ambush for the night moth** satisfy these requirements? By the look of it, the answer is negative; but this answer can be greatly modified. While all the behaviors of Max the cat have been observed by me and my wife Aviva several times, **the ambush of the night moth** was observed only by me and once only. This would appear to be a one-off anecdotal phenomenon, so it does not meet the requirement of publicity and repeatability. However, in principle, so I believe, an experimental setup can be designed in the laboratory that will imitate **the ambush of the night moth** and will allow a series of experiments on this subject. (Actually, I once discussed this matter with my friend Dr. Richard Schuster, who also has a special and interesting cat, named Darwin, but in practical terms we did nothing, because our laboratories at the University of Haifa are not designed for cats but for rats.) Furthermore, it is true that I did not observe Max ambushing a night moth a second time, but I did observe him, as we shall see later, working on ambushes for other creatures, for Aviva, and for myself. However, even if we accept that in principle it is possible to investigate **the ambush of the night moth** scientifically, the foregoing account in no way satisfies the demand for objectivity – there is absolutely no doubt that my love for Max makes its appearance and shines out from every line. The question is if it is possible to render out of the above description an objective behavioral description of **the ambush for the night moth.** The answer is yes. This ambush may be described as an episode, an occurrence, based on a chain of events, on a description of a sequence of behaviors:

1. In a posture of standing-sitting, Max watches the moths flying around the porch light and striking it occasionally in their flight;
2. The moths leave the porch and Max snoozes in the armchair;
3. After while a large moth enters the porch, flies around the light, hits it from time to time, makes a descent in its flight and zooms around the table;
4. Max springs up and begins to chase the moth, which eventually ascends and comes to rest upside-down on the ceiling;
5. Max looks for the moth (sniffs, etc.) and finally leaps onto the table, stands-sits under the light, his head raised for several long minutes.

There is no doubt that this description has cleared away most of the subjective accretions of the previous one. Still, there remain in this description terms loaded with subjective-human significance, for example, the verbs 'watch', 'chase', 'look for'. These terms may perhaps be partially replaced by descriptive, observational terms, but this substitution is possible only up to a certain limit. It transpires that theoretical concepts cannot be translated into, or fully expressed by, purely descriptive-observational terms alone. The purpose of the philosophical approach to science, the positivist approach, to free science completely of subjective, vague, metaphysical concepts failed, and it proved impossible to sever theoretical concepts from observational concepts (see the discussion on these matters in Rakover, 1990).

To tackle these problems and others, I shall start with a methodological approach, "equal hypotheses testing", which I shall describe later. Meanwhile, considering these remarks, I suggest paying attention to the following three points. First, although a number of times the language in which I shall describe Max's behavior will not be free of rich expressions, it will always be possible to convert these descriptions (up to a certain limit) into drier language, the language of scientific observation: into a behavioral skeleton.

Second, in the present context I prefer a rather rich account, because most of the observations on Max were based on an interaction between him and me or him and Aviva – an interaction based on mental emotional ties between the cat and ourselves (see the following chapter). To describe Max's behavior in this context as purely motor is, in my view, an improper act that is liable to distort the true account of the episode.

Third, in the case of the episode of the **ambush for the night moth**, in which I was a passive observer only, I decided to present the description on the rich linguistic level, because it was important for me to set out for the reader something of the excitement that gripped me at the sight of Max's behavior and the thoughts that raced through my head. This excitement, it seems in hindsight, apparently stemmed from some preconception that cats are cute creatures that live only by their instincts. In consequence of this episode of the **ambush for the night moth** I began to think of the way that was most suitable to understand the behavior of Max the cat. Here too, an answer was found to the question, What on earth possessed you to stray from the routine of an experimental psychologist, a laboratory man, and to devote time and thought to understanding a domestic cat, a pet?

1.2.2 Scientific explanation and anthropomorphism

Anthropomorphism (as stated, the attribution of higher human cognitive processes to explain the behavior of animals) is mainly associated with the researcher George John Romanes. In 1883 he published a book (among others) entitled *Animal intelligence*, in which he develops and uses the anthropomorphist-anecdotal approach in comparative research of animals. One of the book's chapters is devoted to the cat. This chapter contains a large number of anecdotes intended to attest to the cat's high level of intelligence. A large part of these anecdotes were published in *Nature* in the section 'Letters to the Editor – Intellect in Brutes'. (In this section many interesting observations were published on the behavior of various animals. Here it is worth adding that the editor of the journal noted that he was not responsible for the views expressed in correspondence with *Nature.*) Here are some examples described by Romanes in his book. Mr. Bidie from the Government Museum of Madras reports the following anecdote (*Nature* 1879, vol. XX, May 29, p. 96). For two months during which Mr. Bidie was away from his home, two young gentlemen resided in it who displayed a bad attitude to his cats. One cat bore kittens, which she carefully hid behind the shelves in the library. Immediately on Mr. Bidie's return the cat returned her kittens to a corner in his room. Mr. Bidie writes:

> I do not think I have heard of a more remarkable instance of reasoning and affectionate confidence in an animal … The train of reasoning seems to have been as follows: 'Now that my master has returned there is no risk of the kittens being injured by the two young savages …, so I will take them out for my protector to see and admire, and keep them in the corner in which all my former pets have been nursed in safety'.

Mr. Thos. B. Groves published in *Nature* 1879, vol. XX, July 24, p. 291 an account of the behavior of a cat before a mirror. The first time the cat saw itself in the mirror it tried at first to fight the reflection, and then ran behind the mirror. When all its efforts came to naught it tried with its forepaw to feel the mirror from behind, while it looked at its image reflected in the mirror. When nothing came of this effort either the cat stopped looking in the mirror. (The similarity between this behavior of the British cat and Max's behavior in **cat in the mirror** is striking.)

Several anecdotes concerned cats that set cunning traps for birds. Here is one I like especially. Mr. Greenock, in *Nature*, 1879, vol. XX, June 26, p. 196, describes a female cat that learned to ambush birds that came to eat crumbs tossed out of the window by his friend. One night the crumbs were covered by snow, and

> [o]n looking out next morning my friend observed Puss busily engaged scratching away the snow. … saw her take crumbs up from the cleaned space and lay them one after another on the snow. After doing this she retired behind the shrubs to await further developments.

This act was observed several times, and following the success of the hunt Greenock's friend decided to stop scattering crumbs from the window.

A number of anecdotes concern the cat's ability to open doors, a fairly well known phenomenon. Romanes, who examines the difference between door-opening by a human and door-opening done by a cat, concludes thus:

> Hence we can only conclude that the cats in such cases have a very definitive idea as to the mechanical properties of a door; they know that to make it open, even when unlatched, it requires to be *pushed* – a very different thing from trying to imitate any particular action which they may see to be performed for the same purpose by man. (p. 421)

I believe that these examples are enough – they represent the spirit of things and show how Romanes realized his main scientific goal: to give examples of intelligent behavior of animals in support of the evolutionary (Darwinian) claim that there is a continuum of intelligence between the animal and the human (e.g., Boring, 1950; Romanes, 1977/1883).

This approach, as stated, was fiercely attacked especially for its confusing the phenomenon with its meaning, and unjustifiably ascribing to animals higher cognitive mechanisms characteristic of humans. In 1894 C. Lloyd Morgan wrote his book *An introduction to comparative psychology* in which in chapter 3, by virtue of a profound discussion of the mind of the other as different from our own mind, he formulates a principle that defends comparative psychology against hasty and non-critical use of the anecdotal approach; it is known today as 'Lloyd Morgan's canon':

> In no case may we interpret an action as the outcome of the exercise of a higher psychological faculty, if it can be interpreted as the outcome of the exercise of one which stands lower in the psychological scale. (p. 53)

What Morgan suggests in this canon is not a prohibition against the use of analogy from human cognitive processes to cognitive processes in an animal, but a warning against rash use of this analogy: "We are logically bound not to assume the existence of these higher faculties until good reasons shall have been shown for such existence" (p. 59).

I stress this point because in the spirit of the times of experimental psychology and behaviorism (see Epstein, 1998) this canon acquired a strict meaning, that animals (and humans) could not be posited to have an explanation for behavior predicated on a higher process, such as recognition, consciousness, or free will, but one explicable by means of a lower process, such as reflexes, instincts, or innate and automatic cognitive processes (and see Thomas, 1998 for an inappropriate explanation of Morgan's canon).

With the rise of cognitive psychology and the burgeoning of research on consciousness, the outlook on the use of anthropomorphism and anecdotes changed. In 1976 Donald Griffin published his book *The question of animal awareness: Evolutionary continuity of mental experience*, and coined the term 'cognitive ethology', which describes a research field that connects 'cognitive sciences' to 'ethology,' an area concerned with mental states such as knowledge, intention, and recognition, in animals. This change sparked extremely severe criticism, and many researchers are unwilling to accept Griffin's approach – cognitive ethology itself, and of course anthropomorphic explanations. (See, for example, a collection of articles on anthropomorphism and anecdotes in Mitchell, Thompson & Miles, 1997a. The article by Bekoff & Allen in that collection broadly reviews a range of different approaches to cognitive ethology, from those that reject to those that approve.) Based on the book by Mitchell, Thompson and Miles, I have chosen to concentrate on three methodological critiques of the use of anthropomorphism as an explanation for the behavior of animals (see Davis, 1997; de Waal, 1997; Mitchell, Thompson & Miles, 1997b).

1. *Anti-anthropomorphism*: Instead of explaining animals' behavior by attribution of higher human processes (consciousness, awareness, free will, intentions), it is better to use simpler and more mechanistic processes such as reflexes, instincts, associative and automatic learning, and evolutionary adaptation to the environment. This directive is actually an expression of the behaviorist principle of simplicity (and it is evidently similar in several respects to Morgan's canon).
2. *Faulty logic*: The logical basis of attribution of human cognitive process to animals is faulty. The analogy between ourselves and the animal is this: (a) when I think, I behave in a certain way; (b) an animal (Max) behaves in a certain way like me, therefore he thinks like me. This conclusion is wrong and is called 'affirming the consequence.' For example, Max can behave in this certain way for entirely other reasons unconnected to conscious thinking.
3. *Unconscious processes*: Not only have we not yet managed to understand higher processes in humans, there are also strong arguments that these processes (consciousness) are not required for an explanation of a wide range of human behavior, for example, motor behavior, attention, speech, and creativity (e.g., Velmans, 1991).

These three critiques are linked to the following issues in animal research: anthropomorphism and anecdotes, use of explanations in everyday psychology (folk psychology explanations such as 'David ran away because he was scared'), the impossibility of observing processes in the mind of another (the 'other mind problem'), and uncertainty as to the methodological ability of field research to shed light on higher processes in animals. Some of these problems (such as the other mind problem, everyday explanations) I shall discuss again and again in the book, but here I concentrate on an attempt to offer methodological solutions to problems of the anecdotal observation, anthropomorphism, and the question of the use of explanations of everyday psychology in cases of conducting observations similar to those I made of Max's behavior.

My answer to these critiques is this: first, I cannot accept critique 3 (unconscious processes); I reject it for reasons I shall detail shortly. Second I partly accept critiques 1 and 2 but basically I maintain that the methodology of 'equal hypothesis testing' bypasses and avoids these critiques.

Critique 3 (unconscious processes): Velmans (1991) argues that in a large number of behaviors, for example, speech, consciousness of the behavior occurs when the behavior is implemented or usually after it. I become aware of the sentence I have spoken after the sounds of the words have emerged from my mouth. I did not say to myself beforehand what I said, but I said what I said without preparing this sentence in advance in my mind. From examples of this kind Velmans suggests that consciousness has negligible importance in processing information, and that unconscious processes are those that mediate between the stimulus perceived by the individual and the individual's response to this stimulus. What Velmans suggests is the following general formula: Behavior = f(stimuli, unconscious cognitive processes).

I responded to this article by Velmans by designing a thought experiment that I called 'the mental-pool thought experiment.' On its basis I argued that Velmans' reply in the negative to the question 'Is processing of human information conscious?' is right only in part (Rakover, 1996). In my opinion, processes mediating between the stimulus and the response are partly unconscious and partly conscious: at the start information is processed unconsciously, but then part of the information enters consciousness, so it cannot be said that consciousness has no causal status. What I propose, then, is the general formula: Behavior = f(stimuli, unconscious & conscious cognitive processes). (LeDoux, 1996, in his book *The emotional brain* reviewed ample literature showing that conscious processes are based on unconscious processes. I don't dispute this, but as we shall see later, I do argue that consciousness is an important factor in explaining a large part of behavior.)

This is the place to note that Velmans' attempt to show that consciousness has no causal status in the explanation of behavior is not the first in psychology. In 1977 Nisbett and Wilson published a famous article in which they argued that analysis of many experiments in social psychology and in decision making demonstrated that participants had no consciousness of the cognitive processes responsible for their behavior. This argument was based, among other things, on the finding that the explanations given by participants for their behavior were less accurate than alternative explanations suggested by the experimenters. In other words, if cognitive process X is responsible for the behavior of the participant in a given experiment, and if, despite the assumption that the participant has exclusive introspective access to the cognitive processes taking place in his brain, the participant's explanation for his behavior is wrong, it is because the assumption is wrong and the subject has no consciousness of cognitive process X.

In an article in reply, which I published in 1983, I showed (Rakover 1983a) that Nisbett and Wilson's considerations were at fault. I pointed out that the participant probably does have access to cognitive processes despite his wrong explanation. I argued that the participant suggests hypotheses about his behavior relying on his introspection, and that these hypotheses are liable to be wrong. In contrast to the experimenter, who knows very well what the cause of the behavior is (for he has designed the experiment), the participant is found to be in a state of vague knowledge, because he has to test which out of all the cognitive processes happening in his head at a given time is the process responsible for his behavior. Therefore, it is incorrect to suggest that the participant has no ability to be conscious of cognitive processes, and the opposite may be proposed: despite the participant's ability to be conscious of cognitive processes happening in his brain, he is liable to fail in his task of finding from among all of them the one correct cognitive process. That is, my argument rests on the scientific process of testing hypotheses. The participant in an experiment raises hypotheses explaining his behavior based on inner introspective information. And just as in science the scientist is liable to raise incorrect hypotheses because of wrong information based on biased and inaccurate observations, so is the participant-scientist liable to err in inner observations in his introspection. And just as it is not correct to claim that scientists are blind and incapable of conducting observations because their explanations are erroneous, so is it incorrect to suggest that the participant is blind to what takes place in his own mind.

Critiques 1 and 2 (anti-anthropomorphism and faulty logic): The directive-prohibition in critique 1 is avoidance of giving anthropomorphic explanations and instead to propose mechanistic explanations, that is, explanations that are given for material phenomena, explanations acceptable in the natural sciences, for example, explanations of behavior by appeal to reflexes and instincts. (It is of interest to note that Looren de Jong, 2003, suggests that the mechanistic explanation should be used only when this explanation adds something more than the mentalistic explanation.) The anti-anthropomorphic directive is based on the fear of making a type-1 error (false alarm), that is, of giving an explanation based on the assumption that an animal (such as Max) is endowed with human mental qualities, while in fact he (Max) is not so endowed. But in my opinion this directive commits a type-2 error (miss), that is, giving an explanation based on the assumption that an animal (such as Max) is not endowed with human mental qualities but with mechanistic qualities, while in fact he (Max) is indeed endowed with these mental qualities. I shall term this tendency to propose mechanistic explanations for behavior of animals "mechanomorphism". The meaning of the term anthropomorphism is 'formed like a man' (Epstein, 1998, p. 71). In the same way, I propose that the meaning of the term mechanomorphism be 'formed like a machine'. (Some time later I found that Griffin, 2001, also used this term, and according to him the person who coined it was Crist, 1999.) Just as anthropomorphist explanations are liable to err, so, I maintain, are mechanomorphist explanations liable to err.

Griffin (2001) distinguishes two kinds of mistakes: in one kind the mistake is linked to the very possibility that animals are endowed with conscious processes, and in the other kind it is connected to the content of the consciousness that we ascribe to an animal. In this connection, we may have no choice but to raise the hypothesis that the nature of the animal's consciousness is like that of our own. But here we have to recall the well known article of Nagel (1974) who asks: "What is it like to be a Bat?" to grasp how hard it is to imagine the nature of a bat's consciousness (assuming, of course, that bats too are endowed with consciousness).

A similar directive-prohibition may be suggested regarding the use of anecdotal events as behavior attesting that Max is endowed with a certain mental quality. For example, a type-1 error is committed if we accept the episode of the **ambush for the night moth** as attesting that Max is endowed with the mental quality of drawing conclusions, when in fact Max is not endowed in this way; and a type-2 (mechanomorphist) error is committed if we decide that this episode does not attest that Max is endowed with this mental quality, when in fact Max is so endowed. (See a similar discussion of this matter in Lehman, 1997; Silverman, 1997.) It is possible that this directive really reflects the researcher's philosophical position: if you are positivist-behaviorist in your basic outlook you will tend to accept the prohibition of critique 1. But if your outlook is less strict and you accept the philosophical criticism of positivism and the cognitive criticisms of behaviorism, you will tend to reject this prohibition and be fearful of making a sweeping type-2 error.

1.2.3 A methodological proposal: Equal hypotheses testing

In this methodology I attempt to propose a solution to the above problem in the framework of accepted scientific methodology, by avoiding prior commitment to the directive-prohibition that appears in critique 1, anti-anthropomorphism, avoidance of trying to justify anthropomorphic or mechanomorphic explanations, and avoidance of the debate on these matters, which it seems is endless (and see discussions on these issues in Lehman, 1997; Silverman, 1997). This proposal has two principal methodological features: testing of hypotheses and models, and comparing alternative hypotheses and models.

Testing hypotheses and models: Following Burghardt, (1985, 1991) and Rivas and Burghardt (2002) (and see Bekoff & Allen, 1997), who suggest regarding anthropomorphism as a means of raising hypotheses about the cognitive world of animals, hypotheses that have to be tested according to scientific methodology, I suggest the methodology of 'equal hypotheses testing'. The moment I suggest an explanation for Max's behavior by appeal to mental qualities, this is not an unequivocal statement that Max is endowed with qualities partly similar to human mental qualities, but a kind of presentation of a hypothesis alone. When I say that Max is afraid, I ascribe to Max a theoretical concept connected on the one hand to the experience of fear, whose properties are partly similar to human experiences of fear, and supported on the other hand by several behaviors such as flattening his ears backward, tensing his body, howling, and running into a concealed corner. If indeed the hypothesis succeeds in explaining the anecdotes, it is *supported* but not verified. If it does not, it is *disconfirmed* and it has to be modified or replaced. (As can be seen, this methodology in essence is no other than the application of the Popperian approach, so-called falsificationism, to the present case (see Popper, 1934/1972).)

Some of the theoretical concepts in current psychology are based on what we, human beings, describe as our private experience, private behavior. For example, when I say that a sudden loud noise frightens me, I mean that I have a private conscious unpleasant experience that my pulse races, that my body reacts by trembling or freezing, etc. The concept of fear, then, is grounded on the one hand in behavior that can be observed by anyone – public behavior (heart rate, the trembling reaction, freezing), and on the other hand it is grounded in an experience that I alone can observe through what is called 'introspection' – private behavior. And when I ascribe to Max the concept of fear as an explanation for his behavior, according to this approach I raise a twofold hypothesis: on the one hand I predict that in the situation of a loud and sudden noise Max will react with public behavior (flatten his ears, crouch, run away), and on the other hand *I hypothesize that in Max a conscious unpleasant experience arises that is responsible for his public behavior.* This is an important point so I shall describe it in detail. (Here it is worth noting that the discussion of theories about the relations among behavior, physiology, feeling, and cognition, such as the James-Lange and Cannon-Bard theories, is beyond the aims of this book. See, for example, on these matters Lazarus, 1991; LeDoux, 1996.)

The commonsense argument is this: without the private conscious experience – the experience of fear or pain, I will not react in the way I do normally to stimuli of fear and pain. If the experience of pain did not arise in me, I would not feel the painful fact that my great toe is broken, and without the experience of fear I would not blink even if I saw a roaring lion about to pounce on me. It is a fact that only blocking the experience of pain by a suitable injection allows the dentist to give me root canal treatment. Now when I ascribe to Max the concept of pain as private behavior, as a conscious private behavior, *I assume as a hypothesis only that Max is endowed with qualities similar to the experience of my private pain, and that this experience is likely to cause him to behave in a way similar to mine.* If this hypothesis is supported by anecdotal descriptions, the hypothesis is borne out and supported by the observations; if not, I am obliged to discard this hypothesis or change it. Let us look at the example of the episode of **musical indifference**.

Let us hypothesize that Max enjoys classical music as I do, that classical music arouses in him the experience of beauty and elevation as it does in me. Is this hypothesis supported by the anecdotal observation? The answer is negative. The episode slaps us in the face: classical music (regardless of composer, kind of music, opera, a violin or piano concerto, a mass, a requiem, or Rossini's *Silken Ladder*) does not move Max – he just doesn't respond in any way, doesn't even prick up his ears toward the TV set from which these marvelous sounds emerge. According to Morris (1997) cats apparently do show a certain sensitivity to music, especially when a specific sound is linked to motherly, sexual, or defensive behavior. This **musical indifference** thus refutes the hypothesis on Max's musicality – the cat is deaf to Mozart, Rossini, and Verdi (how disappointing). (This example substantiates the fact that not all explanations for anecdotes are ad hoc. Moreover, according to the present approach, an ad hoc explanation can be modified and turn into an interesting hypothesis, which can be tested by some new anecdotes.)

Comparison of alternative hypotheses and models: Rivas and Burhardt (2002) write that anthropomorphism is a tendency so deep-rooted and harmful that is has to be fought like the devil: "Anthropomorphism is like Satan in the Bible – it comes in many guises and can catch you unawares!" (p. 15). They counsel overcoming this deep-seated inclination with the help of an experiment to understand the animal's world from its own viewpoint, that is, to step into the animal's shoes. This is a useful guide, which takes the researcher's attention away from his world to the animal's, and to all the scientific information accumulated on it; I have applied it myself in an effort to understand Max. For example, in such an effort I always took into account that his sensory system of hearing, smell, taste, and vision (in certain conditions) is incomparably better than mine. (Furthermore, in connection with a cat's behavior I was helped by articles published in the professional literature and the following books: Morris (1986, 1997); Taylor (1986); Bradshaw (2002); Leyhausen (1979); Tabor (1997).)

In methodological terms I propose overcoming this anthropomorphist tendency, and also the mechanomorphist one, by a comparison of every anthropomorphist explanation with a mechanomorphist one, and vice versa. This comparison, it transpires, is not new and it is worth noting that Romanes (1977/1883) also, about 120 years ago, suggested something similar when he wrote about the learning criterion and the distinction between mentalist and mechanist behavior:

> The criterion of mind … is as follows: Does the organism learn to make new adjustments, or to modify old ones, in accordance with the results of its own individual experience? If it does so, the fact cannot be due merely to reflex action … for it is impossible that heredity can have provided in advance for innovations upon, or alterations of, its machinery during the lifetime of a particular individual. (pp 4-5)

Only if we find that the mechnomorphist explanation is less efficient than the anthropomorphist will we prefer the latter to the former. Methodologically, then, this proposal places the two kinds of explanation on the same level, that of the empirical test. Despite the argument that different theories have different theoretical-empirical weight, I suggest that what ultimately determines which explanation we take is the result of the empirical test. (Here it might be stressed that the methodology for accepting and rejecting hypotheses is far more complicated. On additional considerations for accepting/rejecting hypotheses, and on the importance of the theoretical-empirical weight of hypotheses and theories, see chapter 4, and Rakover 1990; 2003.)

1.2.4 Mechanistic explanations and mentalistic explanations

What are the justifications for the requirement of comparing an anthropomorphist explanation with a mechanomorphist explanation? The answer is this. I maintain that the behavior of animals (including people) is explicable by means of two *explanation approaches*, two basic categories of explanations, which exhaust all explanations known to me: the *mentalistic explanation* and the *mechanistic explanation*. To the best of my knowledge, explanations for the behavior of animals may be classified in the first category or in the second. Therefore, when we propose a mentalistic explanation for an animal's behavior (an anthropomorphist explanation) we must compare it with a mechanistic explanation, in order to test which of these two possible explanations is more efficient.

I include in the *mentalistic explanation approach* everyday explanations that ordinary people give for the behavior of other people. For example, David went to Tel Aviv because he wanted to meet Ruth there. These explanations, which are called everyday, or folk psychology, suggest that in a given situation public behavior (David went to Tel Aviv) is a function of private behavior (David wanted to meet Ruth). That is, the scheme of the mentalistic explanation for the behavior is: public behavior=f(private behavior, situation), where I include in private behavior the feelings, emotions, wishes, beliefs, and conscious and subjective thoughts of the individual.

The mentalistic explanation rests on concepts which, also to the best of my knowledge, cannot be explained in material terms. This is not a tenet, but a hypothesis based on contemporary knowledge (which I consider especially in chapter 8), knowledge which I understand as attesting that conscious mentalistic states and processes still cannot be explained in physical and neurophysiological terms.

Mentalistic explanations, as I said above, are no more than explanatory hypotheses that have to pass the empirical test. For example, Max flattened his ears and ran off to hide behind the big mirror because he was afraid of a strange person who entered our apartment. The mailman ran out of the yard because he was afraid of the barking dog. In both cases the explanation is an everyday one; in both cases I assume that the explanatory concept – fear – is connected on the one hand to public behavior and on the other to an unpleasant private experience. In both cases these explanations are nothing but hypotheses for empirical testing.

I include in the *mechanistic explanation approach* all the explanations for the behavior of animals acceptable in the natural sciences. These include explanations by means of physical, chemical, physiological, genetic, and evolutionary factors, and also explanations grounded in analogy with the computer, such as classical models or neural networks. (I believe that the computer is nothing but a machine that performs manipulations on physical units or symbols, and I do not believe that one day a miracle will take place and a very complex computer will suddenly develop conscious behavior like us humans. On this, see the following chapters.) These, then, are explanations for public material phenomena grounded in material factors or causes. That is, the scheme of the mechanistic explanation for behavior is: public behavior=f(material factors, situation), where these material factors are represented by terms linked directly or indirectly to public observations. These explanations may be of different kinds (causal, probabilistic, dependent on laws of nature or on certain processes and mechanisms), as long as they are based on material terms.

Mechanistic explanations may be formulated by means of formal systems (logic, mathematics, computer language) and the theoretical terms that appear in them can be reliably and validly connected to empirical observations. By contrast, mentalistic explanations are extremely hard to describe with these formal systems, and it is difficult to connect the terms that appear in them (e.g., will, desire, intention, purpose, and belief) reliably and validly to empirical observations. (It is worth adding that the question of how to link theoretical concepts that appear in these two explanatory approaches to empirical observations is complex, and one that I shall not go into here. See on this subject chapter 4, and Michell, 1990; 1999; Rakover, 1990.)

This distinction between these two explanatory approaches raises several problems, to which I shall return in the following chapters of the book. Given a certain behavioral episode, it is hard to decide if this is mechanistic behavior – subject to a mechanistic explanation, or mentalistic behavior – subject to a mentalistic explanation, a decision that in many cases requires an empirical investigation. To illustrate this we may study the following case. David stepped on the tail of a dog, which emitted a howl and bit him. One explanation for the dog's behavior addresses its feeling of pain and fear: the sense of pain-fear is that which made the dog respond with a defensive reaction (howling, biting); another explanation for the dog's violent response is like the 'pain-elicited aggression' reflex. Several experiments on rats have demonstrated 'shock-induced aggression' similar to reflexive behavior: rats given an electric shock begin to react aggressively to each other (e.g., biting). As the shock grows more powerful, the aggression increases (Ulrich 1966). As we shall see later (chapter 4) the fear-aggression reactions in the cat have an inherited-evolutionary background characteristic of cats.

This problem of matching an appropriate explanation (mechanistic – mentalistic) to a given behavior arises from the very fact that the behavioral phenomenon is complex and multi-dimensional, and requires use of two kinds of explanations: mechanistic and mentalistic (a situation I shall deal with in the following chapters). Other factors impeding a solution to this problem are as follows.

Private, mental behavior as explaining and as explained. The mentalistic explanation assumes that a large part of everyday public behavior is explained by inner mental factors, that is, private behavior. But this private behavior does not constitute only an explanatory factor, a cause, for another behavior, public behavior, but in itself constitutes a behavioral phenomenon that requires explanation. When David scurried away from the barking dog, we explained this behavior of flight by addressing the experience of fear that overwhelmed David. But how may we explain the phenomenon of fear in itself? How may we explain the fact that while David sped like an arrow from the dog, Uri not only did not run away but actually behaved so aggressively that the dog ran away from him? Why is David such a coward? And now, if we are able to answer this question, if we explain the phenomenon of fear, will not, ultimately, this explanation for private behavior be in essence a mechanistic (neurophysiological) explanation? And if the answer is affirmative, it will be possible to suggest that every behavior (objective and subjective) may, ultimately, be subject to a mechanistic explanation, hence to suggest further that mentalistic explanation may be reduced to mechanistic explanation. I do not think so. In my view, despite endless attempts to propose a mechanistic explanation for all behavioral phenomena, we still have not been able to explain consciousness and awareness by means of the methodology developed in the natural sciences, and we still have not been able to reduce mental experiences such as will and belief to material factors (and again, see discussion on these matters in the following chapters, especially chapter 8).

Mind and body. As may be seen, these two explanatory approaches are associated with the classic mind-body dichotomy, where mechanistic explanations are applied to material phenomena (body) and mentalistic explanations are given to phenomena that entail a complex (and incomprehensible) interaction between consciousness and the brain, between mind and body. I have no doubt that this statement will spark objections in a large number of readers, arising from the belief that prevails today in a large number of researchers that in the end everything will have a material explanation. I doubt it. My outlook, then, is close in spirit to "explanatory dualism", on which I shall expand in the final chapter, chapter 9; more precisely, my outlook is what I call "methodological dualism". On the one hand, methodological dualism does not demand acceptance of the ontological distinctions between mind and matter, but on the other hand it proposes the distinction between these two explanatory approaches, and argues that behavior of animals (including humans) requires use of mechanistic and mentalistic explanations alike (see on this Rakover 1997, and as noted chapter 9).

Breakdown of behavior and matching of explanations. I assume that these two explanatory approaches cannot be bridged, nor can one be reduced to the other, so a unified (mechanistic) explanation cannot be proposed for all behaviors. This assumption raises the problem of a match between the kind of explanation and the kind of behavior, especially in light of the fact, as stated, that most behaviors of animals are not simple and uni-dimensional but are complex and elaborate, and may be broken down into different behavioral components. How, then, can we know how to distinguish behavioral components subject to mentalistic explanations from those subject to mechanistic explanations?

Answering this question calls for a broad methodological development, which will be conducted in the following chapters (see especially chapters 5 and 6). Here I wish to consider the importance of the help that the professional literature is likely to extend to us in our attempt at an answer. The information accumulating in the literature is based on an empirical examination of hypotheses and theories about the behavior of animals. This general knowledge of the behavior of cats will serve us in the present case as follows. (a) Every anecdotal observation of Max will be resolved into its behavioral components, and (b) each and every behavioral component will be set against the general knowledge of this behavior in cats, with the aim of answering the question of whether this behavioral component of Max is subject to a mechanistic or a mentalistic explanation.

In a large number of cases explanations for the general behavior of a cat are mechanistic, that is, neurophysiological, genetic, and evolutionary, for example, reflexes, such as widening/narrowing of the pupil to a light stimulus; the cat's righting of its body in a fall (a reflex I found in Max too. See Taylor 1986); instinctive behavior in mating or hunting. This last concept, instinct, has undergone many theoretical modifications, from a narrow notion, entirely mechanistic, referring to behavior controlled by inherited neurophysiological mechanisms and released, from latent to active, in a specific condition of stimuli, to a wide concept, treating behavior as an innate-learned behavior, that is, behavior influenced both by inherited evolutionary factors and by learning processes, and characteristic of certain species (such as the cat) (see a discussion on this matter in Barnett, 1998; Haraway & Maples, 1998; Hogan, 1998; Gariepy, 1998). We may take the example of hunting behavior. This is based on an infrastructure of innate behavioral components adapted by means of learning processes to various environmental conditions (e.g., different kinds of fields and woods). It is thus not possible to explain this behavior mechanistically or mentalistically alone, and each behavioral component has to be matched with the kind of explanation suited to it – a research process that will be furthered by the knowledge that has accumulated in the literature.

Explanatory approaches and choice of hypotheses. These two explanatory approaches guide us, each in its own way, to the choice of relevant hypotheses on the explanation of a given behavior. Theoretically, a large number of hypotheses exist for every behavior, and the question is which hypothesis shall we choose in order to explain this behavior. The mentalistic explanatory approach leads us to propose hypotheses based on our personal experience, to bring up everyday explanations by means of which we understand other people, explanations which, as stated, are called folk psychology. For example, I have learned that the notion of pain refers to internal, private, unpleasant feelings, and to a certain kind of public behavior. I have likewise learned that other people who respond with a similar public behavior to mine report that they are in a state of pain. As a result, I assume that they have private behavior similar to mine. In the same way, I assume that animals that evince behavior similar to the public pain behavior of humans will have private behavior (the experience of pain) similar to mine and to that of all other people.

As may be seen, these inferences are based on analogies (see Lehman, 1997; Romanes, 1977/1883). These inferences are not necessarily valid. For example, from the fact that the ass is similar to the human in a large number of properties (both of them breathe, eat, defecate, urinate, sleep, etc.) it does not necessarily follow that the human has four legs. Still, because we customarily ascribe to other humans a private world similar to our own, we have a strong tendency to humanize animals such as dogs and cats, and to explain their behavior similarly to the way we explain the behavior of other humans, by attribution of private behavior.

By contrast, the mechanistic explanatory approach leads us to propose hypotheses based on the immense success of science. That is, to seek hypotheses that will explain animals' behavior in a way similar to the explanations that appear in the natural sciences, by turning to mechanistic explanations. Here too the application is one of analogy. For example, because everything, ultimately, is matter (even the brain is nothing but matter), the argument is put forward that animals' behavior is to be explained by the same kind of successful explanation prevalent in the natural sciences, the mechanistic explanation. Here too the analogy is likely to lead us to a wrong inference. For example, from the fact that the ass does not move, and does not bray despite cruel blows rained down on its back, it does not necessary follow that it does not feel a conscious sensation of pain, and that it is nothing but a machine devoid of feeling and consciousness whose behavior is best explained by explanations characteristic of machine systems.

The creation of explanatory hypotheses, then, is likely to stem from various considerations and justifications: from hypotheses that are based on personal subjective knowledge to those based on scientific knowledge. These considerations have enormous importance for getting an answer to these questions: which hypotheses should we choose to explain a given behavior, and which hypotheses should we choose for research, in order to investigate them by means of additional observations. However, the moment we have chosen the mechanistic or the mentalistic hypothesis for research, at that moment, according to the present approach of equal hypotheses testing, *the methodological status of these explanatory hypotheses is equal: both are bound by the procedure of testing of hypotheses followed in science.*

According to the present methodological approach, then, the researcher is permitted to ascribe to humans and to animals different mental states as a hypothesis – one that explains the observations the most efficiently. The justification for this, in that case, does not lie in the thesis of "incorrigibility", in the hypothesis that humans have direct, infallible observations of mental states, that humans cannot err in respect of their mental states (a thesis very hard to test, for example, in Max the cat), but in the ability of this hypothesis to explain in the best way the complex of observations we have. I do not believe that the incorrigibility thesis is correct, mainly because conducting an internal observation is connected to a complex system of hypotheses and observations, of mental, cognitive, and neurophysiological processes, all of which are subject to error (and see summaries and discussions in Brook & Stainton, 2001; Rakover, 1983a, 1990).

CHAPTER 2

Anecdotes and the methodology of testing hypotheses

The purpose of this chapter is twofold: (a) to propose a methodology of how to construct hypotheses from anecdotes and how to put these hypotheses to the empirical test. First, a hypothesis is proposed as the best explanation for a given anecdote. Then this explanation, this specific hypothesis, is made a general hypothesis by its generalization across these variables: animals, similar situations, similar responses, and time. Finally, the general hypothesis is tested through comparison of the predictions deriving from it with other anecdotal episodes. (b) To propose a criterion, the "principle of new application", whereby a mentalistic explanation is to be matched to a behavior characterized by achieving different goals through the same response, or achieving the same goal through different responses. These methods will be applied to testing two hypotheses on Max's behavior: a fear hypothesis, and the hypothesis that Max ambushed the moth in order to enjoy himself. The latter hypothesis is mentalistic, and is tested by being compared with the mechanistic hypothesis that Max's behavior is innate hunting behavior.

I am quite sure that a computer is just a machine, so much so that one day I typed on my computer a quotation of the philosopher John W. Day and the computer at once leapt up and wrote: Have a good day, and I immediately thought: What a good soul.

In the previous chapter I described a methodology that makes it possible to avoid the tendency to explain animals' behavior by means of anthropomorphism and even the opposite tendency, mechanomorphism. The basic idea is to place these two kinds of explanation on the same methodological level of empirical test.

The question that arises now is double: first, how does one construct hypotheses, theories, from anecdotes? Second, how does one put these hypotheses, theories, to the empirical test by means of anecdotes? The answers to these questions will appear in this chapter in three parts. (I shall use the terms hypotheses and theories to signify the same scientific meaning, even though the term theory refers to a wider and more complex scientific structure than hypothesis.)

First I shall describe briefly Max's living space, his topography (namely the Rakover family apartment: see Figure 2.1), because it was here that the observations of Max's behavior were made, conducted by myself and by Aviva (my wife). I call these observations 'anecdotes' or 'episodes', because they are not observations of the accepted kind in laboratory and field experiments. Most of them are descriptions of observations of interactions between Max and me and Aviva, that is, episodes in which Aviva and I were participant observers.

Then I shall move on to a discussion of the methodology whereby hypotheses are formulated out of anecdotes that occurred in Max's living space, and tested. Finally I shall apply this methodology to this question: did Max indeed plan the ambush for the night moth in the **ambush for the night moth** episode described earlier?

2.1 The living space of Max the cat

My daughter Shelly brought Max to us as a birthday present for me in 1995, when he was four months old. This is a Himalayan cat, bought in Haifa from a family that specialized in breeding this variety. Max has a body structure typical of the Persian cat: short legs, long fur of a golden hue, with grey patches, black extremities (nose, ears, paws, and tail), and blue eyes. Max has the nature typical of the Himalayan cat: he is curious, takes the initiative, and is very attached to us, especially to Aviva (see Taylor, 1986). Max was not gelded, and he lost his virginity at about the age of three with a female Siamese cat that stayed with us for some five days during which we were regaled with a rich repertoire of love songs by the couple. The affair reached its climax as I almost stepped on the pair of lovers, who chose to conduct their amours precisely between my legs while I was shaving in the bathroom.

Max's acclimatization to our third-floor apartment was for the first months accompanied by little escapades. In the first weeks we were busy searching for the small kitten, who got lost everywhere in the apartment: behind the washing machine, the refrigerator, the stove, in any closet with its door left open for a moment. Once he jumped from our kitchen window onto the small kitchen roof of the apartment below, and from there to the branches of a tree, from which he dropped to the yard. My son Omer, who heard the thump of the kitten landing on the roof below, shot like an arrow down to the yard and brought back the little adventurer. About three months after we thought he had got used to our apartment Max slipped out of the open front door, and disappeared. We searched for him for two days, and even posted notices in the neighborhood shops. Then, on the third day, when Aviva was asking neighbors at the entrance to the elevator on the first floor if they had seen our cat, Max stepped out of the electricity box in the wall of the second-floor corridor, and with heartbreaking meows and trembling legs he went down the stairs to her. It turned out that he had been hiding in this box for two days without food or water, had got into a state of distress, and when he recognized Aviva's voice he emerged. This might be one of the reasons for Max's close attachment to Aviva. Other reasons for this firm connection, which do not contradict this explanation, are these: Aviva's female smell (Max, as stated, is not gelded) and the fact that it is Aviva who takes care of the cat, gives him food and drink, grooms him, plays with him and spoils him, and is his companion for most of the hours of the day.

Throughout his nine years with us Max has eaten one kind of food, the 'science diet', and drinks water, in a set place in the kitchen. Max eats and drinks by his free will. He does his business in a litter box with its floor covered with sand, placed on the kitchen porch (see Figure 2.1). Max has regularly had his standard shots, and apart from one time when he suffered ringworm and three times diarrhea, he is completely healthy.

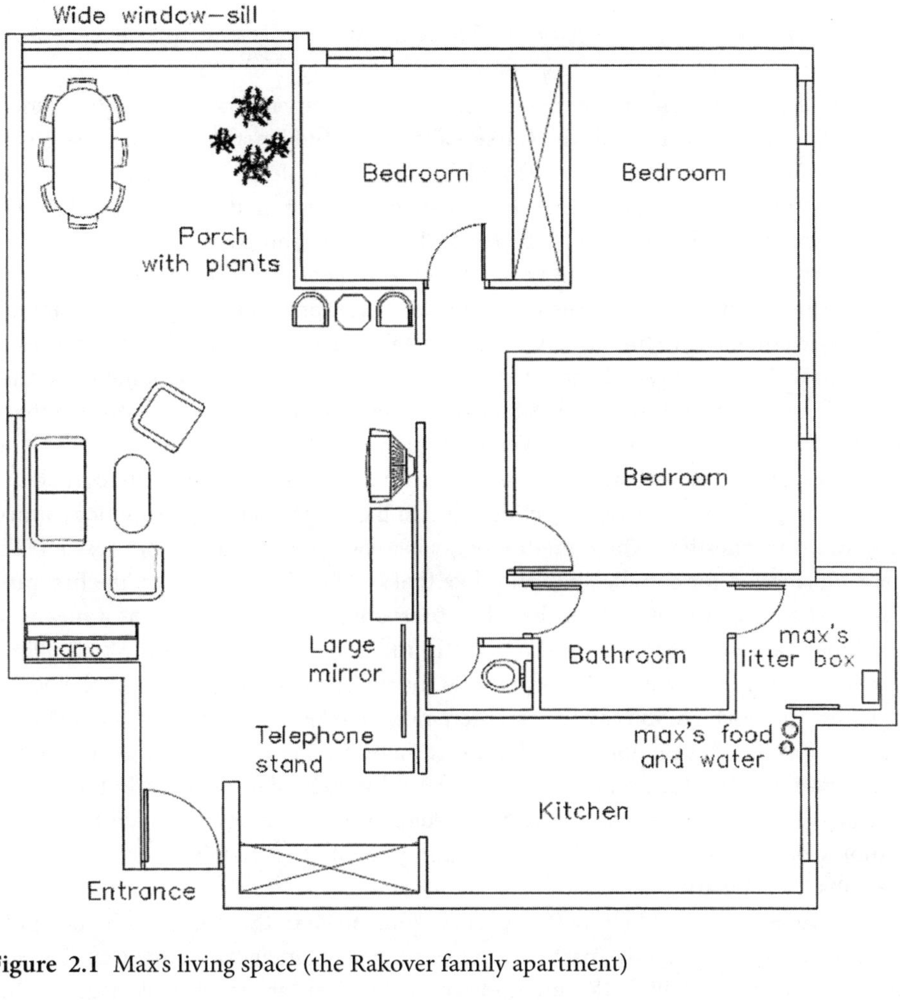

Figure 2.1 Max's living space (the Rakover family apartment)

Max sleeps anywhere in the lounge or the kitchen. Mostly he sleeps on the sofa in front of the television, but several times he has been found sleeping on the chairs around the table on the lounge porch and on the table itself (on warm summer days). Although the entire apartment is part of the cat's living space, we have forbidden him (by closing the door and scolding) to go into the bedrooms. These prohibitions are infringed now and then, and reveal very interesting behavior, as we shall see. We took away the carpets to make the apartment easier to clean. Next to the bathroom we fixed a short pole with a mat wrapped around it for him to sharpen his claws on, but sad to say he does this on the chairs and armchairs in the lounge, in addition to the padded pole. Despite the damage our feelings toward Max have been, and remain warm, cosseting, and loving.

2.2 Pros and cons of observations of Max the cat

In methodological terms, research on Max the cat (observations and their theoretical explanation) may be classified with the so-called case-study method (see discussions in Bekoff & Allen, 1997; Bromley, 1986; Gomm, Hammersley & Foster, 2000; Hamel, 1993). Furthermore, as will become clear soon, the research on the cat may be classified as participant observation, where the researcher, the person conducting the observations, participates in the life of the respondent. (Many of the observations of Max the cat describe an interaction between him and us – Aviva and Sam.) Case study is among the research methods in the social sciences, and as such it serves important purposes in common with other methods: to propose an explanation for the phenomenon under study by interweaving it with the web of similar observations in the appropriate theoretical framework, and sometimes to encourage the development of a new theory. (And see the discussion and debates on these and other matters in the above bibliography.)

By comparison, the research on Max the cat has another and new function, aimed to answer this question: which methodology is to be used in order to propose an efficient explanation for complex behavior like that of Max the cat? That is, the first purpose of the present work is to utilize the observations on Max the cat as a means of developing a new methodological approach, methodological dualism, which allows proper use, from the scientific viewpoint, of mentalistic and mechanistic explanations together. The research on Max the cat, then, may be taken as a test case, one that realizes the methodological approach developed here. So the present goal is not just to interweave the case study with a broad theoretical-empirical framework, but to use the case study to develop new scientific game-rules, a methodology intended to address complex behaviors (based on behavioral components that require mechanistic and mentalistic explanations alike).

At the same time, I believe, the observations on Max the cat in his living space (namely the Rakover family apartment) have several advantages and qualities that will help greatly in developing the methodological dualism approach suggested in this book. Compared with controlled laboratory and field experiments, the observations of Max's behavior are wanting: they have no laboratory control, it is very hard to repeat the observations as is done in the laboratory, or to adapt them to the usual design of statistical research in laboratory and field experiments (and see, on research on a single participant, Bekoff & Allen, 1997; Bromley, 1986; Gomm, Hammersley & Foster, 2000; Hamel, 1993). In addition, while in laboratory and field experiments observations of the subject's responses are recorded at the time the behavior takes place, Max's anecdotes are mostly recorded from memory. This fact allows the influence of explanatory tendencies to enter the recording of the anecdotes.

Nevertheless, I believe that the observations of Max's behavior for nine years in his natural living space have supplied us with information that is extremely difficult to acquire from experiments. (See similar views in Bekoff & Allen, 1997; Griffin, 1981, 2001. See also opposing views in Allen & Bekoff, 1995; Heyes & Dickinson, 1990, 1995. Personally I maintain that research of mental processes is so difficult and complicated that there is no point to a methodological belief that salvation will come only from experiments and field studies.) The main reasons for the relative advantages of observations of the behavior of Max the cat are the following.

A. Free choice: Observations on Max are observations of whole and natural behavior, and are not observations of one arbitrary response performed by an animal in the laboratory, such as pressing on a lever in a Skinner box to get a grain of food, or to escape/avoid an electric shock. By contrast, the episode of the **ambush for the night moth** describes a succession of behaviors done naturally in a certain situation that arose in the cat's natural living space. I believe that in this situation the cat is likely to display and express very interesting cognitive processes that are hard to observe in the laboratory. In particular I mean behavior of *spontaneous free choice*, which in my opinion differs from behavior of *constrained free choice*.

I assume that part of an animal's behavior is an expression of its free will, expressed in non-coerced choice, when coerced choice is exemplified by mechanistic behavior such as reflexes and instincts (see discussion in chapter 3, and in Barnett, 1998; McFee, 2000). For example, in laboratory experiments, such as lever pressing in which a hungry rat learns to press a lever in order to obtain reinforcement (food), the range of the animals responses is limited to two categories: one response that brings reinforcement (pressing the lever) and all the other responses that do not bring reinforcement. If the rat presses the lever, the response is considered an indicator of learning; if it does not, the non-pressing, which includes various responses and non-response (the rat simply does nothing), is considered an indicator of non-learning. But in the case of the latter category (non-pressing the lever) the possibility also exists that the rat, as a living creature, simply does not want to respond. In this and similar experiments, then, one does not check whether or not non-response is done out of free will, simply because by definition non-response is considered non-learning.

Laboratory experiments are therefore based on *constrained free choice* because in them the possibility of not choosing between two alternatives is eliminated. The animal is constrained to respond in a way that interests the experimenter, as in the lever-pressing experiment in which the animal is starved and caged in the experimental apparatus, which drastically cuts down the repertoire of natural responses (otherwise it would be impossible to run the experiment).

But with observations on Max in his natural living space the possibility of spontaneous free choice immediately presents itself. For example, when Aviva calls Max he responds to her in most cases, changes course, and comes to her. Yet in a number of cases, while turning his head to her, he chooses not to approach her and continues on his way (**non-change of direction**). Although the choice of going to Aviva is very nicely rewarded by a pleasant stroking, there is no coercion here to choose this response, and the cat, despite turning his head to Aviva, has chosen, spontaneously and freely, to continue moving in the same direction as before. This kind of behavior does not exist in the lever-pressing experiment because, as stated, the experimental conditions constrain the animal to behave according to what is expected of it. (Yet note here that experiments of this kind supply the most important kinds of information on such questions as what are the causes of fast learning as against slow learning.)

This distinction between *spontaneous free choice* and *constrained free choice* gives rise to two important problems: once again the problem of type-1 error, and the problem of an ad hoc explanation (an explanation after the fact). Attribution of free choice to animals is of course anthropomorphist. Perhaps animals do not have consciousness and free will: 'I could have chosen otherwise'. However, as I noted in the previous chapter, this assumption about free will is nothing other than a kind of hypothesis about the consciousness of Max the cat. This hypothesis was tested here by the episode of **non-change of direction**. But here one may ask: How do you know that the explanation of this episode too is not flawed by anthropomorphism? My answer is that although this question is apt, and although I do not have a decisive answer, I will say that the empirical test of a number of other episodes attesting to free choice supports the attribution of free will to the cat more than the attribution of the mechanomorphist hypothesis, that is, its perception as a mechanistic system. (See the following chapter on indicators of free will.)

An experimental psychologist is likely to argue that the explanation for non-response in a Skinner box in a lever-pressing experiment as the exercise of free choice not to respond is nothing but an ad hoc explanation, which may be ascribed to every case in which participants fail to solve the problem placed before them. Accordingly, it is always possible to put forward the explanation that this rat has not learned to press the lever not because it is plain stupid but because it chose of its own free will not to respond. Or this student failed in an exam not because he is dumb but because he decided of his own free will not to study and to fail. Clearly, this criticism is possible. It is very likely that the rat is stupid and it is very likely that the student likewise is stupid – these are legitimate suppositions, in my opinion, just like the hypothesis of free will. These hypotheses are testable by means of more empirical observations, and in that way may be released from the ad hoc prison. For example, if it turns out that the student is manifestly stupid the free-will hypothesis will be greatly weakened. And if it transpires that our rat, in other tests, achieves dazzling success and displays high intelligence, would this not reinforce the hypothesis that this rat decided of its own free will that it was sick and tired of the experiment, and it wasn't going to cooperate with the experimenter any more?

B. Communication: In Max's living space almost the only creatures whose company he keeps every day are Aviva and myself. We are the creatures with whom Max has to communicate, not only to get food and water (which, as I mentioned earlier, are supplied to him regularly and in a fixed place), but also to form an emotional, social tie. So Max had to invent means of communication in order to get our attention, and the development of these means of communication, as Griffin suggests in chapter 10 of his book *The question of animal awareness* (Griffin, 1981, p. 149), may constitute a possible window on the minds of animals. Such development, to the best of my knowledge and experience, simply does not happen in laboratory research because there rats are treated as entirely passive creatures, whose whole function is reduced to giving one response (or two responses) to a programmed and defined set of stimuli. By this I do not mean that laboratory research is not able to construct a set of experiments that will supply us with important information on the animal's mental processes: the reverse is the case. All I wish to state is that conducting observations in Max's living space is likely to yield interesting information about the cat's internal world because in this space Max has to initiate ways of communication with us – a situation that does not exist in the laboratory. (However, as stated, there are different views on the subject. While Heyes & Dickinson, 1990, underline the importance of the experiment as a way to vindicate attribution of intentionality to animals, Allen & Bekoff, 1995, suggest that field observations are essential. And see the response of Heyes & Dickinson, 1995, to Allen & Bekoff's critique.)

C. Simple field observations and experiments: A large part of the observations of Max's behavior were, as noted, participant observations, namely based on an interaction between Max and ourselves (Aviva and me), on games with the cat, and on simple experiments in which I changed a bit of Max's physical living space (e.g., prevention of passage by closing doors). In these field experiments it was possible to watch how the cat handled a problem he came up against. In not one of these experiments did we use punishment or withhold food from the cat.

It transpires, then, that the book will address a certain kind of behavior of Max the cat. It will not try to describe and explain physiological, visual, learned responses, as these are regularly investigated in the laboratory. I shall deal with the cat's *daily behavior*, as performed by Max. This is the behavior of an adult cat which in part is instinctive and in part learned, mentalistic, and adapted to the cat's living space.

D. Behavioral structure and its parts: As I noted above, laboratory and field experiments (e.g., learning to press on a lever) center on one response out of a collection of responses that constitute a whole behavioral structure. This may cause several problems, which we shall exemplify by discussing play behavior in kittens (e.g., Barrett & Bateson, 1978; Bateson & Young, 1981; Burghardet, 1998; Caro, 1981). Bateson and

Young found that on being weaned, kittens intensified their play behavior, which was measured by recording the frequency of appearance of several play responses, such as "object contact" defined as "Each pat with a paw making contact with an object (particularly the toy dog, ball or stuffed rabbit) and each bite of these objects" (p. 174). Anyone who has watched cats handling a ball, for example, sees at once that part of this behavior is touching the ball with a forepaw. So I shall call the structure of responses connected to activity with a ball "ball activity", and contact with the ball by a paw "ball touch", and I shall raise two questions:

First, on what grounds is it determined that the behavioral structure of ball activity is indeed play behavior?

Second, is it not possible to ascribe to an isolated ball-touch response different interpretations, unconnected with the play interpretation?

To clarify these two questions let us look at the following two response schemes, where R is response:

Episode A: [R1→R2→(R3)→R4→R5]

Episode B: [R6→R7→(R3)→R8→R9]

where the symbol → means that one response follows another response.

Episode A as a whole is interpreted as play: the cat approaches at a run, touches the ball with its paw (R3), the ball begins to roll, and the cat chases it;

Episode B as a whole is explained as curiosity: the cat approaches crouching, touches the ball with its paw (R3), the ball begins to roll, and the cat, lying on its belly, watches the ball's movement.

On what grounds did we interpret the first episode as a game and the second as curiosity? My answer is this: if we do not treat both these episodes as purely behavioral descriptions, as motor occurrences only, but as behavioral structures carrying meaning (play and curiosity), then in fact we are interpreting them analogously to our own behavior. And as soon as we call episode A play, we are applying anthropomorphism. Barrett and Bateson (1978) write, "Definitions of play are notoriously difficult even though the observer has a strong subjective sense of certainty about when an animal is playing" (p 107). Moreover, when we conduct observations of (R3), on ball touch alone, and we treat (R3) as a response displaying the behavioral structure of episode A, we are liable to commit a double error: the humanization of the cat's response, and the specific interpretation of this response, since (R3) may also represent episode B, namely curiosity. This possible confusion is resolved through reference to the context in which the specific response (R3) appears. For example, Caro (1981) conducted observations of several isolated responses, and classified them into two different behavioral categories according to their behavioral context: play or hunt. Hence, the interpretation of an isolated response greatly depends on the meaning of an everyday behavior in which this response is embedded.

Does not this example, which shows that it is very hard (if not impossible) to avoid regarding animals' behavior from our point of view, weaken my approach of 'equal hypotheses testing' that I developed in the last chapter? My answer is that it does not, because the process of hypothesis testing (mentalistic or mechanistic) is insensitive to the way in which it is created and sensitive only to the question of matching the predictions derived from the hypotheses with the empirical observations (and see a wide-ranging discussion of this in chapter 4).

2.3 Construction and testing of hypotheses from anecdotes

Following Cronbach (1975), who writes: "When we give proper weight to local conditions, any generalization is a working hypothesis, not a conclusion" (p. 125), I decided to treat the explanation of a behavioral episode as a source for constructing hypotheses that may be put to the empirical test. The question, of course, is how can an explanation of a given phenomenon, an anecdote, be turned into a working hypothesis? My suggestion is the following.

The general scheme for building hypotheses and theories, and testing them, is this: an individual's behavior is a function of the situation to which the individual is subject and of the theory that connects, in an explanatory way, the behavior and the situation: Behavior = f(Situation, Theory). This scheme helps us to understand what is done in an experiment, and what should be done methodologically with anecdotal observations.

The *experiment* is based on an attempt to answer the following question:

Behavior(?) = f(Situation, Theory).

That is, given a situation and a theory, the experiment asks what behavior will be obtained. In other words, the question established by the experiment is this: given a certain situation and a certain theory, will the predicted behavior indeed be obtained?

By contrast, in *anecdotal* observation an attempt is made to answer the following question:

Behavior = f[Situation, Theory(?)].

That is, given behavior in a certain situation, what theory must we construct such that it will connect in the most efficient explanatory way the behavior and the situation? In the present case, then, the behavior in a certain situation is given us, and what we are looking for is the best theory. For example, in the episode of the **ambush of the night moth** Max's behavior in certain situations that arose in his living space (the flight of the moths around the light at night, etc.) is given. What we lack is the explanation for his behavior – the ambush for the moth. How, then, does one construct and test theories from anecdotes?

In the *first stage* we try to suggest, that is, to apply to the present case or to devise, the best hypotheses that will explain the given episode, such as the **ambush of the night moth.** This is a known inductive technique called 'inference to the best explanation' (see Josephson & Josephson, 1994; Lipton, 1991, 2001a, 2001b). By means of this technique we may propose for the **ambush of the night moth** answers to the following kind of questions: starting from the mentalistic viewpoint, what are the mental processes that acted in Max's consciousness to cause him to behave in the way he did? And starting from the mechanistic viewpoint, what are the cognitive-neurophysiological processes that acted in Max's brain to cause him to behave in the way he did?

In the *second stage* we must take care (a) to remove the explanation for the anecdote from a methodological status of an ad hoc explanation, and (b) to choose out of several alternative explanations (that have been given for the anecdote) the most effective one. To realize these two aims we must test these explanations by means of other relevant anecdotes. This we do by turning the *particular explanation* for Max's behavior (**ambush for the night moth**) into a *general explanatory hypothesis*, namely by generalizing the particular explanation across the following variables: *animals* (all Himalayan cats, cats in general, and so on, including humans); *similar situations; similar responses;* and *time.*

In the *third stage*, after we have turned the particular explanation for the anecdote into a general explanation, we can test this explanatory hypothesis on various animals, in similar situations, by similar responses, and at different times. If we find that in these tests our predictions are indeed confirmed, that is, we see that the predictions are equal to the behavioral observations, we may confirm one explanation as more efficient than alternative explanations.

Since all our observations were of one subject, Max, it is worth expanding and describing empirically how to test the general explanatory hypothesis (which we shall denote H_G) in similar situations and over similar responses. Schematically, this proposed process follows the so-called Hypothetico-Deductive (H-D) method (see Glymour, 1980; Rakover, 1990, 2003; Salmon, 1967):

From H_G and situation 1 (denoted S_1) we derive a prediction that Max will react with response R_1;

From (H_G and S_2) we derive R_2;

From (H_G and S_3) we derive R_3;

From (H_G and S_i) we derive R_i, where i denotes any suitable stimulus and response.

Now, all that remains is to find anecdotes that occurred in a certain situation (S_1 or S_2 or S_i) and test if indeed the predicted response (R_1 or R_2 or R_i) matches the response that appears in a given anecdote. If a match is found H_G is confirmed; if not it is refuted. (Here it should be stressed again that contrary to anecdotal observations, in a laboratory experiment the desired situation is constructed, and the occurrence of the predicted response is tested for.)

As a simple example, let us look at the episode of the **escape response**: when our neighbor Nurit walked into our house on a visit, accompanied by her big dog Snow, Max scuttled away fast behind the big mirror in the lounge.

The immediate mentalistic explanation is this: Max ran away from the frightening dog and found a hiding-place behind the mirror. To make this explanation testable, we shall turn it into a general hypothesis:

The *fear hypothesis*: For ALL ANIMALS, at ALL TIMES, when they enter a THREATENING SITUATION they respond with FEAR RESPONSES.

The words in capitals represent general concepts, variables, to which different values can be assigned. Empirically, there is hardly any problem with the concepts of animals and time, but several problems arise with the two remaining concepts. It is not clear how to define observationally a stimulus that arouses fear and fearful responses, because to a large extent what arouses fear depends on the individual herself. A certain stimulus is liable to arouse a fearful response in one but not in another, and a response thought to be of fear in one is not thought to be so in another. Still, fairly wide agreement exists that painful stimuli, sudden and strong stimuli, and stimuli that through learning processes have become connected to the stimuli mentioned above – all these stimuli are thought to arouse fear. Similarly, there is agreement that a startle response, freezing (crouching), fleeing, and fighting, which are largely innate responses (a roaring lion – who will not fear him?) and in part are learnt (through classical or instrumental conditioning), are fear responses.

Now, after we have connected the theoretical concepts that appear in the fear hypothesis to situations and responses of certain kinds, let us test in Max if this hypothesis holds empirically for other anecdotes. Was the reason for Max's running away indeed fear? Would Max respond to new threatening situations with responses thought to be fearful responses? We shall look at two more episodes.

Defense response: After Max disappeared behind the mirror (see the **escape response** episode) Snow began sniffing around the mirror. Suddenly the front half of Max's body emerged from behind the mirror and his head approached the dog's head to within about thirty centimeters. At once Max arched his body and hissed. The upshot was that Snow got very frightened and ran out of our apartment.

It is reasonable to suggest that in this threatening situation, with the close presence of the dog, Max entered into a state of fear accompanied by aggression, and responded as he did. Morris (1986, 1997) describes similar responses in cats.

Visit to the vet: When Max goes for treatment to the vet he is placed in a special box. Max vehemently objects to getting into the box, and he vociferates the entire way to the vet's clinic. But once there Max refuses to get out of the box. On the way back home from the vet's he does not make a sound. At home he gets out of the box posthaste and runs away to a concealed corner, under the table on the porch, or behind the big mirror.

It is very reasonable to assume that Max's responses in this case are mainly fear responses, but the nature of the stimulus that aroused the fear is not clear. If we assume that the fear-arousing stimulus is the *box itself*, it will be hard to explain the set of Max's responses. The box, as a threatening stimulus, a negative stimulus, may explain his objection to getting into it in the Rakover house, his loud mewing on the way to the vet, and the eagerness with which he jumps out of the box when he is brought home from the vet's. But the box as a negative stimulus, in itself, cannot explain his responses at the vet's clinic: his refusal to get out of the box, and the speed at which he gets back into it after his treatment. To explain this behavior we must assume that in addition to the box as a negative stimulus, the "vet place", is a kind of negative stimulus whose threatening power is greater than the power of the box.

Yet it is still hard to explain why Max is silent on the journey after his visit to the vet, while he set up a racket all the way there. I think that to explain the different pre- and post-vet behavior we must assume that Max has learned everything associated with the episode of the visit to the vet, and conclude from this that the cat knew he was being taken to the vet, and therefore he objected and mewed, and he knew that the treatment at the vet's was over, so he hurried back into the box and did not mew on the way home. This hypothesis, that Max responds to what he has learned, also explains how the box has turned into a negative stimulus. At first Max displayed curiosity about it and sniffed at it, but after we put him inside it and took him for his visit to the vet his attitude to the box changed in accordance with the context of this episode: before the treatment at the vet's the box is a threatening stimulus and after it its positive value rises. Back home he quickly gets out of it, because the value of the home, his living space, is far more positive than the box, which is connected with unpleasant events.

As stated above, the fear hypothesis cannot be tested across the variable of animals, simply because all the observations were conducted on Max alone. The observations on Max, as noted, are included methodologically in research done on one participant, a case study (see discussion in Bekoff & Allen, 1997; Bromley, 1986; Gomm, Hammersley & Foster, 2000; Hamel, 1993). Although I am unable to discuss the complex of problems linked to the subject, in the present context I would like to consider very briefly the following questions:

a. Are the anecdotes accidental observations?
b. Do the anecdotes describe behavior that appears in other cats (animals)?
c. Is the theory given to a certain anecdote ad hoc?
d. Is the theory exclusive for Max?

I do not uphold the possibility that Max's behavior is chance and idiosyncratic because a large part of it is described in the literature, which constitutes in the present case a standard reference point for evaluating the cat's behavior. Nor do I maintain that the hypotheses that I use to explain Max's behavior are ad hoc and idiosyncratic, because these hypotheses are based either on everyday explanations or on other explanations that have received empirical support at least partially. And, as stated, these hypotheses have been tested empirically by their applicability to other anecdotes of Max the cat.

2.4 Test of the hypothesis that Max ambushed the night moth for his amusement

In the previous chapter I interpreted the episode of the **ambush of the night moth** as behavior that attests to high cognitive activity: Max, I argued, drew a logical conclusion from earlier knowledge and behaved accordingly: he waylaid the night moth that was likely to resume its flying around the porch light. Is not this interpretation anthropomorphist? May not this behavior be explained by means of the hunting or playing instinct? To answer these questions I resort to the literature describing observations of cats regarding general and characteristic behavioral patterns for states of hunting and play (see Taylor, 1986; Morris, 1986, 1997; Barnett, 1998; Burghardt, 1998; Caro, 1981; Leyhausen, 1979; Tabor, 1997). I compare Max's behavior in the episode of the **ambush for the night moth** with these behavioral patterns, and examine if it is possible to regard his as an individual example of the general characteristic pattern. If it is, I shall see Max's behavior as no more than another empirical observation that supports the general explanatory hypothesis that in states of hunting and play cats behave in a certain way as described in the literature, that is, I shall see Max's behavior as an individual case of instinctive behavior.

This comparison is not simple, because the behavior we are dealing with is complex and made up of quite a number of behavioral components. In fact, every behavior, episode, anecdote, is not a one-dimensional response to one stimulus, a simple reflex, but an organized collection of sub-behaviors. As a result, the process of comparing the characteristic description (in the literature) with the anecdotal observation itself is complex, and is based, among other things, on the following procedure, which I shall call "exhausting the mechanistic explanation". (This procedure is based on the inductive principle of John Stuart Mill called "the canon (method) of residues" (J. S. Mill, 1865). According to the residue principle, if a given phenomenon is wholly divisible, for example, into three sub-phenomena, A, B, C, when phenomenon A is explained by factor *a*, phenomenon B by factor *b*, then the residue – C – will be explained by another factor, factor *c*.)

Procedure of exhausting the mechanistic explanation: I assume that the description of a given behavior (of a cat) may be divided into several descriptions of behavioral components, to a certain extent independence of each other. First, the sequence of actions of the cat in the hunt, for example, is not necessarily a complete mechanical chain of reactions but a sequence that may vary according to the situations of the hunt, the behavior of the prey, and the physiological-mental state of the cat itself. Second, even though every behavioral component is testable theoretically/empirically in itself, this component has an important function in the overall behavior (e.g., Rakover, 2003). Therefore, one must examine (a) if every behavioral component is explicable mechanistically, (b) if the organization of all the behaviors is mechanistic, and (c) if the purpose of the use of the entire behavior also is given to a mechanistic explanation (e.g., a genetic, evolutionary explanation).

If every single component can be explained by means of a mechanistic explanation, and if it is possible to explain mechanistically the organization of the parts, it will be possible to propose a whole mechanistic explanation for the entire behavior, that is, it will be possible to exhaust the mechanistic explanation to the end. However, if we find that not all the components are interpretable mechanistically, or if it transpires that their organization does not accord with a mechanistic explanation, a mentalistic explanation will be required, and therefore the entire phenomenon will be explained by a dual explanation, one based on two kinds of explanation: a mechanistic and a mentalistic. Furthermore, even if it emerges that the purpose of the use of the entire behavior is not given to a mechanistic explanation, we will need a dual explanation.

Let us look, for example, at the mechanistic explanation of the operation of a flashlight: when we press on the switch it lights up. The explanation is based on (a) the explanations of the parts of the flashlight – bulb, battery, electrical wires, and switch, and by the appropriate physical and chemical theory; and (b) by the arrangement of the flashlight parts according to the theory of electricity (completing and breaking the electrical circuit). Hence one may deduce that the phenomenon of illumination of the flashlight is explicable entirely mechanistically. But this is not the case when one examines the purpose of using the flashlight. The use is not given to a mechanistic explanation because the purpose of the flashlight's use is determined by the person, and because a number of properties of the flashlight, such as its shape, its color, and its weight, also constitute an expression of the user's requirements of the instrument. Consequently, the explanation of the flashlight's operation is wholly mechanistic, but the purpose of its use is not.

Considering these clarifications and the example of the flashlight, I shall discuss the following two questions: is the episode of the **ambush for the night moth**, which is divisible into sub-behaviors, wholly satisfied by a mechanistic explanation? And if it is not, what behaviors, their organization, or their purpose require a mentalistic explanation?

To answer these questions we shall move on to a short description of the cat's hunting and play behavior (see descriptions and discussion in Taylor, 1986; Morris, 1986, 1997; Layhausen, 1979; Tabor, 1997). A wild cat, which stalks its prey stealthily, begins to approach it at a crouching run, stops and watches, gets ever closer to the prey from hiding place to hiding place, waylays the prey, and when it is fairly close it moves quickly to the attack, leaps on the prey, pinions it to the ground with its forepaws and stuns it, and when it is in the right position the cat gives it the coup de grace – a bite into the neck that tears the spine. This behavior, which basically is instinctive, has to be learned. Therefore, the cat's young that have not learned from their mother how to hunt and eat the catch find it very hard to perform hunting behavior smoothly and their degree of success is slight.

In contrast to the wild cat, which is an expert hunter, the housecat, which is not adept at this work, tests again and again whether the prey has gone into shock so that it may safely deliver the bite into its neck without itself being hurt and wounded as the prey, a rodent or a bird, may bite, scratch, and lacerate the cat's face. To us these tests look as if the cat is playing with its victim. Indeed, many researchers have found that the hunting behavior is very similar to play behavior, for example, with a ball of paper. Several components of the behavior are common to the hunt and to play; hunt and play are sparked by similar stimuli (fast-moving small objects); and hunger has a similar effect on hunting and playing behavior (Caro, 1981; Hall & Bradshaw, 1998; Hall, Bradshaw & Robison, 2002; Leyhausen, 1979; Tabor, 1997).

Does not Max's behavior in the episode of the **ambush for the night moth** show great similarity to the account of hunting behavior? The answer is affirmative in part. The response of prolonged staring at the moths stimulated in Max by the flying insects' rapid movement (or their shadows skittering over the ceiling and the walls) around the lamp and their crashing into the lampshade, and Max's reaction of chasing the moth that flew around the table, may be explained as part of the hunt/play response aroused in the cat when it discerns fast motion of small creatures or objects. This response may be produced in the cat also by rapid movement of a patch of light cast onto the floor by a flashlight. Max chased after the patch of light in fast circuits and from time to time he leapt at it, thrusting out his two forepaws (the **patch of light game**).

But how may we explain the observation that after the night moth disappeared, and after Max was frustrated in his search for it (a search that appears in the behavior of hunting prey that manages to slip away occasionally), Max jumped onto the table and positioned himself under the lamp with his head raised upwards? Even if we try to explain this as part of hunting behavior (the cat searches for the escaped prey and follows its trail, sniffing without pause), the episode of the **ambush for the night moth** will still retain a fragment of unexplained behavior: Max waited for the moth below the porch light. By this I do not argue that this behavior is special to Max – to the contrary, cats customarily lie in wait. Waylaying is an important part of hunt behavior: the cat comes to a place where the prey customarily comes, and awaits it hidden. Furthermore, Max himself usually lies in wait, especially for Aviva. When Aviva enters her room or the bathroom, Max ambushes her in the archway between the lounge and the passage leading to those rooms. The rear half of his body is in the lounge and the front half is in the passage. There he sits or lies, his head turned and his ears cocked towards the room where Aviva is. The moment she leaves the room Max gets up and goes to her (**ambush for Aviva**).

So it is reasonable to suppose that Max ambushed the moth just as he ambushes Aviva. But if we accept this assumption of the ambush, it is not possible to ignore the following question: on what grounds did Max prepare the ambush? Because readying an ambush requires the use of prior knowledge. Without knowing or assuming that the prey is likely reappear in a certain place, no ambush can be got ready. Max, then, needed this information to prepare ambushes. In the episode of the **ambush for Aviva** Max saw where Aviva went, and also knew, by virtue of long previous experience, that she would come out from that place. But this is not the case with the episode of the **ambush of the night moth**: here Max did not know where the moth had disappeared, so he was bound to deduce, through earlier information, that the moth was likely to resume its flight around the night light. (This information is apparently based on two observations: the moth's flight around the lamp before it descended to fly around the table, and associating this moth with the group of moths that earlier flew around the lamp.) In brief, what I suggest is that the episode of the **ambush of the night moth** calls for a dual explanation: mechanistic (based on the hunting instinct) and mentalistic (based on the use of mental processes).

2.5 Matching a mentalistic explanation to behavior: The Principle of New Application

The use of dual explanations for observations of complex behaviors led me to the following question: for what kind of behavioral components, behavioral states, is the suggestion of a mentalistic explanation called for? The attempt to answer this question produced the formulation of a new principle, which I call the "Principle of New Application," that considers the behavioral element likely to need mentalistic explanations.

The Principle of New Application*: The following behavioral cases are likely to attest to mentalistic activity (intention, belief, purpose, knowledge, awareness, and thinking) and require mentalistic explanations:*

A. *The use of existing behaviors (mainly instinctive) to attain new goals, different from the goals achieved previously by means of these behaviors;*
B. *The use of different behaviors to achieve the same goal.*

What justification underlies this principle? The justification is based on the mentalistic goal-directed scheme: if X has the will (*W*) to achieve a certain goal (*G*), and if X believes that performing a certain action (*A*) will achieve (*G*), it will be logical for X to perform (*A*).

This is a scheme of mentalistic explanation that we apply to *observations* of the following kind: X performed action A and achieved goal G. But in several cases these observations may at first sight be assigned a mentalistic explanation even though they are explained by a mechanistic explanation. For example, a torpedo's behavior seems goal-directed, even though its operating and navigation equipment is wholly mechanistic; the behavior of Max, falling freely with his legs above and his back below, seems mentalistic because in the course of the fall the cat flips over and lands on his four paws, even though this behavior is instinctive, innate, and characteristic of cats. To eliminate errors of this kind and to increase the likelihood of fitting of the mentalistic explanation to this behavior, I propose the Principle of New Application. In my opinion, this principle deals with an area of behavior for which it will be hard to suggest a mechanistic explanation. Why? The answer consists of two parts: first I shall examine how this principle is connected to the mentalistic goal-directed explanation, and then I shall test if it is possible to propose a mechanistic explanation even for these special situations.

Mentalistic explanation: To justify the first part of the Principle of New Application by means of the goal-directed explanation, we shall examine the following two observations:

(a) X performed A to achieve G_1

(b) X performed A to achieve G_2

According to (a) and the goal-driven explanation it may be deduced that: X had $Will_1$ (W_1) to achieve G and X believed that the performance of A would realize his will. Now, if we accept this inference as reasonable, it is also reasonable for us to accept the following inference regarding observation (b): X had W_2 to achieve G_2 and X believed that performing A would realize his will. The difference between observation (a) and observation (b) is that two different goals were achieved by performance of the same action A. This difference requires a parallel change in the Will (W) of the individual, and not in his Belief, because in both cases the individual performed the same action (A), that is, he believed that A would bring about both and G_1 and G_2.

To justify the second part of this principle by means of the goal-directed explanation we shall examine the following two observations:

(c) X performed A_1 and achieved G

(d) X performed A_2 and achieved G

Now, by means of the process of inference similar to the one above, we can infer that X wanted to achieve G and had two beliefs: A_1 would realize his will (observation (c)), and A_2 likewise would realize his will (observation (d)).

Mechanistic explanation: Now I ask, is it possible to propose for both observations (a) and (b) a mechanistic explanation? My answer is that it will be difficult to do so. If we suggest a mechanistic explanation for (a): X performed A to achieve G_1, that is, X is neurophysiologically programmed (like a machine) to perform A in order to achieve G_1, it will be hard to suggest the same explanation for (b): X performed A to achieve G_2, that is, for the observations that X achieved other possible goals (G_2 G_3 …) by means of the same action A, because these goals are not part of the machine's program. To explain the achievement of other goals one must turn to the inner world of X, to his different wills to achieve different goals by means of the same action.

Similarly, it is hard to suggest a mechanistic explanation for the two observations (c) and (d). If X is neurophysiologically programmed (like a machine) to perform A_1 to achieve G, it will be hard to suggest the same explanation for observation (d): X performed A_2 to achieve G, that is, for the observations that X achieved G by means of different actions. To explain the achievement of the same goal by means of different actions, one has to turn to X's inner world, his will to achieve the same goal by means of his belief that different actions will realize his will.

On these grounds it is possible to propose the Principle of New Application, according to which different goals achieved by the same behavior, and the same goal achieved by different responses, attest to an appropriate mental activity (change of will or belief).

In the two episodes of the **ambush for Aviva** and the **ambush for the night moth**, Max demonstrated the use of mainly instinctive behavior in order to achieve new goals. In the former episode Max used hunting/ambush behavior to achieve attention and petting from Aviva. These are entirely different goals from those of the hunt. In the latter episode the explanation is slightly more complicated because the instinctive hypothesis can be put forward: Max's goal in the ambush of the night moth was to hunt and eat the moth. In my view this hypothesis is not supported on account of the following observations and considerations, which sustain the mentalistic hypothesis that Max wanted to amuse himself.

First, I do not believe that Max's intention was to satisfy his hunger by eating the moth. As noted, Max has a 24-hour a day food supply, he is not famished like an alley cat, and does not have to go to any trouble to satisfy his hunger pangs. Furthermore, Max eats only one thing, the special food we buy for him at the pet shop. This is dry, nourishing food, which contains everything a cat needs. He refuses to eat any other food. It transpires that this is not such unusual behavior in housecats, which show reluctance to touch new foods (e.g., Bradshaw et al., 2000).

Dr. Richard Schuster, my good friend at the Psychology department, who researches social behavior in animals, told me that his cat, Darwin, behaved like Max too. One day, feeling kind-hearted, Richard decided to give Darwin a treat, to prepare him a feast for his dinner. He bought a can of expensive and tasty meat, whose smell made Richard's mouth water. He tossed Darwin's regular food into the trash, and spooned onto his plate this cordon bleu dish. He saved the rest of the meat from the can in the fridge for other meals. Richard was certain that Darwin would hurry over to the tasty meal, gobble it down, and lick his lips in sheer bliss. But what did Darwin do? Sniffed the food, turned his feline rear end to it, and went to stretch out on the couch. Alright, Richard thought, maybe he's not all that hungry. The next morning the food still lay on the plate. Okay, Richard said to himself, you'll eat that or stay hungry. I didn't spend a fortune of money on you for nothing. A day passed, and another; the meat in the plate began to change color and go off, and Richard had to throw it out. Darwin didn't eat. He just drank water. Eventually Richard, with a heart-rending sigh, tossed out what was left of the entire can and restored to Darwin his accustomed food.

Secondly, Max is a housecat, whose entire world is limited to the area of our apartment; he has never learned to hunt. Whoever thinks that knowledge of how to hunt and prey is wholly present from birth in the cat's head is making a serious mistake. As stated, cats learn to hunt. An important part of their education for hunting is taught by the mother cat. If the cat does not gain experience or train in stalking and killing prey, it will not chase mice, pounce on birds, or know how to put its hunting instinct into practice. Without learning, the cat's responses will not adapt to the environment, which changes ceaselessly. Learning is what ensures the cat's vast capacity for adaptation to the various theaters of the hunt, which alter without end (see on this, e.g., Morris, 1986; Tabor, 1997).

Well then: what was Max busy with there under the porch light? In my opinion Max wanted to amuse himself with the artful moth, to chase after it again and again, as he was accustomed to do with the ball of paper (**a ball of paper game**). I crumple a sheet of paper and make it into a sphere the size of a ping-pong ball and toss it in front of Max's nose. The cat begins to chase it, jabs it with his forepaws, continues chasing the ball as it rolls under the tables and chairs, jumps on it, traps it between his legs, punches it away, and again chases after the evasive ball. This behavior, which also has been described in the literature cited above, is constructed of a large number of behavioral components of hunting.

Other games too between Max and Aviva and me are composed of behavioral hunting components. For example, we may look at the **game of scratches and biting.** Max is positioned in front of me in his seated-sphinx posture with his two forepaws outstretched. I pass my right hand quickly in front of his eyes, to the sides, forward and back, aiming to tap him lightly on the head with my fingers. For his part, Max tries to strike my hand with his forepaws or to grab it in his mouth. He is fast, and ultimately leaves a few scratches on my arm, which brings the game to an end as I hurry to the bathroom to disinfect them. Moreover, when he sometimes succeeds in getting my hand in his mouth, his bite is light and does not wound.

While this game activates instinctive hunting responses in Max (even though he has never really bitten my hand!) it is hard to see what remnant of instinctive behavior is triggered by the following games. These are games which, I believe, support the hypothesis that Max wants to enjoy himself, just like me.

The armchair: catch. Max likes to play this game with me and with Aviva. When I return from the kitchen with a cup of coffee, planning to sit in my armchair facing the television set, Max beats me to it with a sprint and leaps into the armchair an instant before I am about to sit in it. His body is all tense and his glittering eyes are fixed on me. "Max", I say loudly and jokingly, "What are you doing, you naughty cat, taking my place on purpose?" I place the cup on the table and the moment I stretch my hand out to him he springs out of the armchair and begins to run around, throwing back at me fast looks as if to say, "Catch me – if you can". Now the real fun begins. I jump at Max, who with amazing agility escapes between the chairs and under the tables; I shove the furniture about with scrapes and bangs, thrust out my hand to the tricky cat, who escapes into the kitchen. When I run into the kitchen the cat flees into the bathroom, and from there through the passageway into the lounge, and with a leap into my armchair. And the chase begins again. Just as I reach my arm out to this spectacular cat, he jumps up and runs off. Finally, breathing heavily and laughing heartily, I collapse into the armchair worn out, gulp down some coffee, and declare in surrender, "You win, Max. That's it. I can't move anymore". Max stops, resumes his sphinx-like pose, levels his blue eyes at me for a long staring moment, and then, like one who has accepted the decision – the game is over, begins to lick himself.

While Max is busy with his toilet, I ponder his antics. Can this behavior be explained as instinct? Did Mother Evolution prepare Max to sport with me at **the armchair: catch**? Was this behavior innate, an automatic, involuntary response? I don't think so.

What did Max have to learn and know in order to play this game with me?

First, Max had to learn my habits, that the armchair in front of the TV is mine, where I sit; that when I get up and go to the kitchen and return from there with the cup in my hand I am heading for my armchair in order to go on sitting in it, drinking, and watching television.

Second, Max had to attract my attention so that he could convey to me the message that he wanted to play. And I must admit, the cat passed his wishes on to me with the utmost success. He jumped into the armchair and prevented me from sitting in it, just as I was about to do so, an act that got my attention at once. And the moment our looks met, the moment I stretched out my hand to him, he leapt up and ran off, shooting rapid backward looks at me. Only a fool would not realize that Max was inviting a chase, especially considering that the cat stopped from time to time to check that I had indeed got it, and that I was in motion after him. And indeed I had got it, and skipped after him making laughing and fun noises: "Hey, Max the cat – watch it, the big bear is coming after you, growl growl growl."

Thirdly, Max had to weigh up this complex information in his head, my habitual behavior, the way to pass the message to me about the game in order to achieve his goal – a chasing game. Do you have a better explanation than this? It is hard to imagine that Max, who had grown up with us his whole life, would train in escaping and flight from predators bigger and stronger than he. To the best of my knowledge, the literature has not reported on this.

I believe that additional testimony supporting the explanation that in the episode of **the armchair: chase** Max wished to play and enjoy himself is the game-variant called **the armchair: petting.** This variant attests that Max is able to perform the same behavior in order to achieve a different goal. According to the Principle of New Applications, described above, the behavior attests to high adaptive cognitive activity (awareness and thinking) and requires mentalistic explanations. The explanation for **the armchair: catch** is through attributing to Max the wish to play and have fun, and the explanation for **the armchair: petting** is through attribution of the desire to be petted.

In this latter game, an instant before I make to sit in the armchair Max has jumped into it, flipped over onto his back, and stuck his head back. "Hey Max," I joke, "What do you want, baby, a little petting – stroking and tickling on the belly and under the chin?" And when I stroke him and tickle him, suddenly he grabs my hand between his two front paws without unsheathing his sharp claws, and slightly bites the palm. A touching act. I do not understand how Max is able to prevent his claws from emerging. Every time he stretches his body, sticks out his forepaws, or gets hold of something with his legs, the claws come out in a mechanical and automatic reflex (see Tabor, 1997 on the machinery of extending claws). His canine teeth are long, strong, and sharp. If he wanted he could sink them effortlessly deep into my palm. But he does not. He bites my hand the way we bite our little children, who giggle in pleasure as a sign of love. And frequently, in pleasure and with a wide smile, I have seen Max play in the same way with Aviva. A joyful sight for the eyes.

If the insistent reader still asserts that these last games too, the **armchair: catch** and **petting** games, are in the end based on instinctive, evolutionary components of the hunt, I maintain that the next game, the **tail game**, will attest with a force that can hardly be explained on a mechanistic basis, that indeed Max likes to play and amuse himself for the purpose of enjoyment itself.

As noted, Max likes to play all sorts of games, and it transpires that he catches on very quickly how to play new games. The **tail game** first happened in our large lounge. In front of the television set is a handsome, oval table. To its left stands my armchair, to its right Aviva's armchair, and behind it a sofa, facing the TV. One evening, while I was eating dinner while watching television, Max jumped onto the sofa and came up to me, placed his two forepaws on the table, and with deep curiosity sniffed at my food, over and over. I hadn't the heart to shoo him away and simply chided him, "Max, that's not for you. That's my food. Go on, get off. Sit on the sofa." A little while later the meal was finished. Max turned his back to me, lay on his belly, the upper part of his body resting on the arm of the sofa and his long tail wagging right and left. I wanted to compensate him and I began to play with him. I placed my hand on his tail, pressed it gently to the sofa, and released it. Max moved his tail quickly to the right and to my great astonishment returned it to the place of the pressing a few seconds later. Again I pressed it and let go. This time Max moved his tail to the left and again replaced it on the pressing spot. Again I pressed it, and then again. Press, let go, move the tail, and back to the pressing place. Thus the tail game was created. From the time when this game was invented Max would come to me occasionally, lie on his belly with his wagging tail directed at me, and I never refused to play with him. Sometimes the game took place on the sofa and sometimes on the floor. In summer, when I am sitting in my armchair barefoot, Max lies on the floor and wags his tail directed toward me. I gently press the tail with my bare foot and let go. Max wags his tail right or left, and returns it accurately to the place of the pressing. Over and over, until I get tired. The phenomenon is incredible. How does he remember to put his tail back exactly on the right spot, without turning his head? His mental ability to situate his tail in space is noteworthy. Does this game serve some evolutionary purpose? And is it possible to see the tail game as some element in instinctive behavior? It seems to me that the answers to both these questions are negative.

In sum, what does all that say to us? What can we deduce from the complex of these game behaviors in respect of the **ambush for the night moth**? I believe that this complex shows that parts of Max's complicated behavior require mentalistic interpretations, that is, it is very hard to explain his behavior as a whole by resorting to mechanistic processes and systems alone.

CHAPTER 3

Free will, consciousness, and explanation

The aim of this chapter is to suggest reasons and justifications for explaining the cat's public everyday behavior mentalistically, by appeal to his internal-conscious world. What is the evidence that the cat is endowed with consciousness? The basic idea is to develop a behavioral criterion for free will applicable to humans, and to suggest that the cat's behavior meets this criterion, is performed consciously, like the free-will behavior performed consciously by human beings. This justified attribution of consciousness to the cat allows the use of explanatory terms of the mind, such as intention, will, purpose, and knowledge. In addition, the chapter considers other evidence of consciousness in animals, the possibility that there are levels of consciousness, the possibility of knowing other minds, and the complex relationship of explanations and the individual's awareness of her own behavior.

I went to the cinema of my own free will, because I decided that my wife would choose which film we were going to, just as she determines everything.

In the two preceding chapters I argued that in several cases it was hard to suggest a mechanistic explanation for Max's behavior, so I proposed a mentalistic explanation. In those cases I assumed overtly or covertly that Max acted out of free will. For example, in the interpretation of the episode of the **ambush for the night moth** I suggested that Max waylaid the moth for his amusement, and in the interpretation of the episode of **non-change of direction** I suggested that sometimes Max chose of his free will to obey Aviva's call to go to her and sometimes he decided to continue on his own way.

Furthermore, the very question of whether to explain a given behavior mechanistically or mentalistically, as we shall see at once, touches on the metaphysical question about free will: does a person act out of free will, out of free choice, or is everything perhaps determined causally. This is a weighty and profound matter, on which in the present context I will be able to offer no more than a very brief comment (on this see, e.g., Ekstrom, 2000; Kane, 2002a; McFee, 2000; Smilansky, 2000). In general, three basic approaches exist to the problem of free will. According to the *hard determinist* approach there is no free choice and everything is determined in advance, and the individual's behavior can be explained by means of causal explanations that prevail in the natural sciences. Given a certain situation and a certain natural law, only one sole result-response is determined, so it is impossible to say that in such a state of affairs any other response will be obtained, as we would expect on the assumption of free will and personal responsibility. According to the *libertarian* approach, determinism is untenable because there are people who act out of free will and out of free choice. According to this approach, a human being has the ability to act beyond physical and neurophysiological determinism. According to the *compatibilist* approach, personal responsibility, resting on free will, complies with determinism, for otherwise how may one justify our blaming David for kicking Joey on purpose, but our letting him off for kicking his neurophysiologist as a result of the knee reflex test?

A number of researchers suggests a distinction between a causal-explanation, or an event-explanation, which applies to physical and biological events, phenomena, and a reason-explanation, or an action-explanation, which applies to actions of the person (see, e.g., McFee, 2000; Ginet, 2002). Without going into the unanswerable question of whether these two kinds of explanation are qualitatively different, or into the unending discussion of the many arguments for and against this distinction (arguments that I shall in part discuss later on), I just note the two following things. First, I believe that methodologically an explanation for behavior such as *raising* a hand in greeting by resorting to the individual's intentions differs from an explanation of the motor *movement* of the hand or the *movement* of a billiard ball. In the latter case the causal factor may be observed (movement of the cue striking the ball) independently of the observation of the outcome (the ball's motion), whereas in the case of raising the hand in greeting this movement carries meaning for both the greeter and the greeted, and it is hard to separate the motor movement from its meaning (see similar discussions in Audi, 1993; Ginet, 2002; O'Connor, 2002).

Secondly, I maintain that the following example substantiates this distinction between a causal-explanation and a reason-explanation intuitively. Let us look at the sad case where David commits suicide by leaping off the Eshkol Tower at Haifa University. The causal-explanation of the event of 'the death of David' will center on an account of the neurophysiological process that led David to hurl himself out of the top-story window of the tower. By contrast, the action-explanation for David's deed will concentrate on the mental process responsible for the ending of his life. Given these two explanations, the major question arises: is it possible to translate, or to reduce, the second explanation (which has recourse to the mental process) to the first explanation (which has recourse to the brain, the neurophysiological process)? If the answer is affirmative, the mental explanation may in essence be seen as a causal-explanation. But if the answer is negative we must take the dualist approach, with two different kinds of explanation. As stated, the controversy over this question has not been settled and the debate continues. Personally, I believe that the arguments for the dualist approach are more convincing (see a discussion on this matter later).

From these aspects, it seems to me that the 'methodological dualism' approach (which attempts to determine if Max's behavior is subject to a mechanistic (causal-) or to a mentalistic (reason-) explanation, and proposes that understanding the entire behavior may be based on these two kinds of explanation together) may be regarded as close to the compatibilist approach. On the one hand, I agree that phenomena exist whose most efficient explanation is by means of terms taken from the world of free will, and on the other I agree that phenomena exist that may be explained and predicted by means of theories prevailing in the natural sciences. Still, I think that great differences lie between the compatibilist approach and my own.

A. The philosophy of free will is, at root, a metaphysical, ontological, and ethical discussion, which resounds through Greek philosophy and continues to the present day. This is a ramified discussion, replete with arguments and counter-arguments, which apparently will go on for ever (see wide-ranging reviews in Clarke, 2003; Kane, 2002b). By contrast, methodological dualism is an empirical methodological approach that concentrates on such research questions as these: Is a given behavior of a cat subject to explanation by neurophysiological, genetic, and evolutionary processes, or must other explanations connected to mental processes, that is, private behavior, be resorted to? And can one suggest for a complex behavior, consisting of several sub-behaviors, some of which may be subject to mechanistic explanations and some to mentalistic, a general explanation coherently organized? In other words, methodological dualism is concerned with developing a scientific procedure that will allow the researcher to arrive at reasonable answers to these and similar questions.

B. The question about free will in the cat is also discussed here from the empirical research aspect. For example, does a certain hypothesis, that a given behavior expresses free choice, obtain empirical support? This hypothesis, which ultimately is tried before the court of the empirical test, is constructed by means of different criteria based on philosophical knowledge, scientific knowledge, and everyday psychology. For example, a behavior that is hard to explain by mechanistic explanations is that which may be subject to an interpretation that satisfies the requirements of the Principle of New Application that I posited in the last chapter: when existing behaviors (chiefly instinctive behaviors) are used for the sake of achieving new goals, and when different behaviors are used in order to achieve the same goal.

Here is another example. The philosophy of free will proposes that free will is a condition for behavior that manifests responsibility. This responsibility seems hard to find in the behavior of the cat, as responsibility is in essence normative – do this and nothing else. This behavior is internalized and arouses various internal feelings, such as guilt (conscience). Nevertheless, I will ask: Is behavior associated with responsibility likely to arise in a cat? Is it possible to hypothesize that Max's behavior expresses something in which a feeling of guilt may be discerned, behavior that hints that Max understands that he has infringed a proscription? As we shall see later, there is an indication of this.

C. Methodological dualism assumes that two kinds of explanatory hypotheses, the mechanistic and the mentalistic, may under certain conditions suggest different predictions that are subject to confirmation/refutation. Be the result of the empirical test as it may, according to methodological dualism there is no final decision in favor of this hypothesis or another. Knowledge is provisional. We use certain basic knowledge in which we have great theoretical/empirical confidence, and by its means construct new hypotheses for an empirical test. But even this basic knowledge may be put to a new test when new questions about it arise (on this matter see Rakover, 2003). Scientific research, then, is never-ending, and from this point of view I agree entirely with Popper:

> The game of science is, in principle, without end. He who decides one day that scientific statements do not call for any further test and that they can be regarded as finally verified, retires from the game. (Popper, 1972, p. 53)

In this chapter I shall try to suggest several behavioral indicators of free will. First I will discuss the question of the methodological status of a behavioral indicator, then I will propose several behaviors in Max that are indicators of free will, and finally I will discuss the relationship of free will, consciousness, and explanation.

3.1 The methodological status of indicators of private behavior

The discussion of indicators is based on the following distinction I made in the previous chapter between the question of the experiment and the question of the anecdotal observation:

The *experiment* is based on an attempt to solve the following problem:

Behavior(?) = f(Situation, Theory).

That is, the question that this experiment poses is, given a certain situation (S) and a certain theory (T), will the Behavior, which we derive from S and T, be obtained?

By contrast, with the *anecdotal* observation we try to solve the following problem:

Behavior = f[Situation, Theory(?)]

That is, given a Behavior in a certain situation, the question is, which theory must we construct so that this theory will connect the behavior to the situation in the most efficient explanatory way?

In the present case, the theory we are seeking may acquire different forms, for example, neurophysiological theories, cognitive theories based on analogy to the computer (whether this analogy is built on classic programs or neural networks), or folk theories based on private behavior.

The indicators we are talking about in this chapter are indicators of an animal's private behavior that may explain its public behavior.

A proposal for behavioral indicators: Certain public behaviors B_o that appear in certain situations S are indicators of a certain private behavior B_p when the following conditions are met:

A. There is a theory T_p based on B_p, and empirical observations that support a connection of the kind $B_o = f(S, T_p)$, that is, a family of connections exists between certain situations and certain responses that are explained by T_p;

B. This connection is nothing other than a kind of correlation, association, because S and B_o are likely to indicate B_p, but are not like necessary and sufficient conditions for B_p.

As an example of a family of correlations explained by the concept of fear (B_p), we shall look at three episodes.

Escape response: When our neighbor Nurit visited our apartment accompanied by her big dog Snow, Max ran away quickly behind the large mirror in the lounge. As stated, the mentalistic explanation for this behavior is by means of the concept of fear as representing private behavior: Max ran from the frightening dog and found a hideout behind the mirror.

Defense response: After Max disappeared behind the mirror, Snow began to sniff around it. Suddenly Max stuck his head out from behind the mirror and found himself thirty centimeters away from the head of the dog. At once he arched his body and hissed. The outcome was that Snow was startled, and ran out of our apartment. It is reasonable to suppose that in this threatening situation, in the close presence of the dog, Max went into a situation of fear accompanied by aggressiveness and responded in the way he did, with a defense response.

Startle and freeze response: One day, due to a freak gust of wind, the door slammed shut with a sudden loud bang, which caused in Aviva and in me a fear response expressed in Startle. I noticed that Max too shrank back, and went briefly into a state of immobility.

These three responses – escape, defense, and Startle/freeze – express a situation of fear that arose in Max as a result of three certain (fear-arousing) stimuli: the dog's entering the apartment, the presence of the dog's head close to the cat, a sudden loud noise. In these situations, then, one can see these responses as indicators of Max's private response – the fear response. But here the following question arises: is not this evidence a kind of circular definition? That is, on the one hand I explain the responses to the three situations by means of fear, and on the other hand I use the same situations and responses to infer the private response – fear. Is this circularity destructive? I believe not.

First, here one theory is at issue, which is connected with the concept of fear, and which explains the appearance of the various responses in different situations of stimulus in cats (Max) and other creatures (e.g., humans). Secondly, although to the best of my knowledge there is no decisive definition of a frightening stimulus, in consequence of an inductive collection of relevant observations, frightening stimuli may be characterized by the following qualities: they are stimuli that arouse pain innately, they are strong and sudden stimuli, and they are learned stimuli that have appeared in association with painful situations or the two foregoing kinds of stimuli. Fear responses are those that arise with behavioral stimuli/situations such as: escape, avoidance, Startle, freezing, and defense (attack). These generalizations are, I think, broad enough to break free of the circularity and to establish an empirical-theoretical framework whereby different behaviors that appear in diverse situations of fear in various animals may be explained.

3.2 Indicators of free will in Max the cat

By virtue of the above literature on free will and action (see, e.g., Audi, 1993; Rosenberg, 1988; Stout, 1996), and of everyday logic, I chose a number of behaviors that in human beings are interpreted as expressions of free will, and I applied them to Max's behavior. This approach is undoubtedly anthropomorphist, but I honestly have no idea of what the interpretation of free will is in a cat. All I do know is that in the assumption of free will and personal responsibility in the human, an assumption that accords with the libertarian, compatibilist philosophical outlook, and with everyday psychology, these behaviors are interpreted as free will. This application from the human's behavior to the cat's behavior, as I stressed earlier, is just by way of hypothesis. In this sense the indicators of free will may be regarded as hypotheses on private behavior, that is, on the internal experience that we have a real possibility of free choice and that we are not forced to behave only in a certain way.

These behaviors that allude to free will are not independent of each other, yet despite the partial overlapping I maintain that each and every quality is unique.

Initiative – The individual is the source of the behavior and the behavior springs from her. This behavior is not a kind of reflex or instinct; that is, the individual does not make the same response to the same stimulus at different times. Here it is notable that the start of a new behavior is the end of other behaviors performed by the individual, for example, rising, which concludes sitting or lying. (Some of our behaviors are performed all the time, such as replacement of materials, blood circulation, and activities of heart, lungs, and kidneys. Some of these behaviors we are conscious of, and in a great many of them we are not in a state of consciousness.)

Persistence – The individual controls her behavior and persists in it despite several obstacles that crop up in her path.

Choice – The individual has the possibility of choosing out of a number of behavioral alternatives, including a choice between the possibility of continuing with an existing behavior or changing it. A choice arises also when the individual is confronted with several stimuli and she decides to approach or to distance herself from one of them.

Interruption of the flow of the behavior – The individual interrupts the flow of an initiated behavior without an external factor obliging her to do so, and then resumes with the behavior itself until its conclusion, namely the attainment of her goal.

Infringing prohibitions – The individual tends to violate prohibitions when she believes that she is not under surveillance.

The principle of new application – The individual uses existing behaviors (mainly instinctive behaviors) for the purpose of attaining new goals, and uses different behaviors in order to achieve the same goal.

Here perhaps is the place to suggest a list of kinds of behaviors that Max is able to use to achieve various goals:

1 Scratching with the claws of his forepaws
2 Biting

3 Fixing with looks
4 Loud wailing
5 Movements of ears and of tail
6 Body postures (standing, sitting, crouching, and lying)
7 Body movements (running, jumping).

These behaviors, for the most part, are based on genetic-evolutionary elements and are subject to mechanistic explanations. Yet Max apparently uses them to attain different goals (to create communication between himself and us) and thus he exhibits free will. Analogously, I will suggest that human beings too use mechanical behaviors to achieve various objectives. Clearly, motor movements of my hands and feet are explained by an appeal to the appropriate neurophysiological mechanism. But these movements are harnessed to different goals, according to my free will. And now to the indicators:

Initiative

Jumping onto knees: Max jumps onto my knees or Aviva's without our calling him.

Jumping off knees: Max gets of our knees even when we go on stroking him and are enjoying his presence.

Scratching armchair–knees: While I am watching television in the evening, Max occasionally has the habit of scratching the left side of the seat of the armchair. (By now this spot is torn to pieces.) I turn to him and we exchange looks. Afterwards I respond to him, lean over to him, lift him up, lay him across my knees, stroke him, and Max begins to purr. (In a few cases he gets up and begins to bang against my stomach. This is an instinctive reaction connected with sucking.) In this case, then, Max uses scratching to get petted (stroking, tickling his head, behind his ears).

Scratching throat and jaw: Max loves to have his throat and lower jaw scratched. He indicates that he wants this by raising his head and stretching it backwards.

Combing the fur: Max greatly likes to have his fur combed. Aviva is charged with this job and when she gets back from her morning walk (I'm still sleeping) Max leaps out from wherever he is lazing and runs to her with a loud mewing. She settles herself on the telephone stool and Max at once lies on his belly and then on his side. (Sometimes the fur combing takes place when Max is sitting on the telephone stand.) A number of times, when Aviva was occupied and did not comb him on entering the house, Max attacked the armrest of her armchair with scratches.

Games: As I described earlier, in several cases Max initiated the start of the games, for example the **tail game** (he lies in front of my feet as I sit in the armchair and begins to wag his tail) and the **armchair-catch** or **petting games** (see previous chapter). Another game that Max likes to play is similar to the **armchair-catch game**. In this one, the **chasing game**, Max goads me to chase him when I'm sitting in the armchair. Suddenly he jumps up and thumps the chair's right armrest, and then goes back and flops onto the floor. His eyes meet mine and I discern from his body posture that Max is ready for the chase. When I get up from the armchair the fun begins: Max runs off under tables, the sofa, and the chairs, while I push them around so as to catch him. I call out loudly, "Max, you beautiful cat, here I am to catch you". He slips out of my hands (or between my feet) lithely. Usually the chase ends when Max leaps into my armchair and waits to be petted.

Rolling over onto back: continue petting: Max lies on the floor, I bend over him and stroke and pet him. When I straighten up and am about to go, Max quickly thrusts out his forepaw and slaps it down onto my shoe. When I stroke him again, he takes his paw off my shoe. (This episode raises a thought: does Max connect my shoe and my hand, and treat them as a single entity? If the answer is positive, how can we explain the following behaviors: when I move my finger in front of his face he follows the finger and not any other part of my body. And when I put a pen under the tightly stretched coverlet on the sofa on which Max is lying and move it right and left, Max follows the movement of the pen and hits it with his paw, but not the hand that moves the pen. Perhaps in these cases Max is wholly concentrated on the movement of the stimulus.)

Persistence

Getting off knees: Sometimes, while Max is lying on my knees I notice that he turns his head this way and that, and I sense through my knees that he is about jump of them. Sure enough, he leaps off me. Occasionally I stop him from jumping off, but when I remove my hand he takes the opportunity to jump off. I interpret this behavior as the persistence of will over time.

Jumping onto the sofa: In the evening, as I am glued to the television, I notice Max stepping toward the sofa to the right of the TV set and a little behind the armchair. When Max passes to my right, and changes from walking to readiness leap onto the sofa, I place my right hand on him and stop him from jumping. After about thirty seconds I remove my hand and Max jumps onto the sofa. The interpretation of this case is similar to that of the episode of **getting off knees**.

Licking fur: Several times when Max was busy licking his fur I stopped this activity. Afterwards, when I took my hand off him, Max resumed his prior activity and persisted at it. This episode is more complicated than the foregoing episodes to interpret, because licking behavior is characteristic of cats and is taken as instinctive. Therefore one may suggest the following as an alternative to the interpretation of 'persistence of will': this behavior reflects continuity of the motivational situation that underlies it.

Setting Max on knees: Usually Max obeys Aviva, and when she calls him he comes at once. But some cases occur when he, for example, lying snoozing on the floor, does not answer her call. On these few occasions Aviva rises from her armchair, picks the cat up, and settles him on her knees. But the moment she takes her hand off him Max takes the opportunity to jump off her knees. Again, the interpretation of this case is like that for the two preceding – Max's will persists over time. He didn't want to sit on her knees in the first place, and he does what he wants the moment he can.

Non-change of direction of walk: As stated, when Aviva calls Max he answers her in many cases, alters his behavior (e.g., the direction he is going), and comes to her. But in a number of cases, even though he turns his head toward her while he is still walking, and even stops, he chooses in the end not to go to her and he continues on his way. Although according to Aviva's appeal is well rewarded with petting, the cat continues with his original behavior. The explanation for this behavior, as in the previous cases, suggests that Max continues to satisfy his prior wishes. He had the choice of continuing on his way or going to Aviva (he turned his head in her direction and stopped), and he chose the first option and persisted in its performance.

Non-change of lying: This behavior is very similar to **non-change of direction of walk**. Max lies on the floor, his head between his outspread legs. Aviva calls him and he, even though he turns his ears toward her, does not stir from his place. (In similar cases when I call him, he doesn't even bother to move his ears.)

Choice

Almost every action may be conceptualized between performance and non-performance, that is, performance of other actions. From his aspect, almost all the above episodes are connected with the choice Max made to perform his free will. Yet several episodes exist of a distinct choice by Max.

Preference for Aviva: Max clearly prefers Aviva's company to mine. This preference is expressed in several episodes, which I shall assemble under the present heading. When we are sitting watching television Max sits with face toward her, staring at her. Often he sits near her on the sofa next to her armchair, and brings his nose close to hers. When she calls him the cat mostly sidles up to her dog-like. When she goes to the bathroom, toilet, or her room, he customarily waylays her on the threshold between the lounge and the passage to the bedrooms, the bathroom, and the kitchen (see **ambush for Aviva**). Max usually follows her wherever she goes in our apartment and in many cases he "talks" to Aviva: when he approaches her he emits various kinds of mewing and gurgles.

I don't have the pleasure of such a relationship. Yet this does not mean that Max is indifferent to me. When I get home from work Max usually jumps onto my knees in Aviva's presence (especially after Aviva has lifted him off her knees). After Aviva goes to bed I stay to watch the sumo or boxing shows that I like. At the end of the programs, when I turn off the TV and go to my room, Max suddenly passes me at a run, and under my feet rolls over on his back, fixing me with his eyes (**sleep: running-rolling over on the back**). The meaning of his action is so clear that not even once do I not bend down and pet him for long moments.

The interpretation of this collection of episodes is clear: Max has a positive attitude to me, but with Aviva it is far more positive. So Max prefers Aviva to me, but at night, when I am left alone, he prefers to stay with me rather than be on his own.

Places for sleeping and napping: Max generally varies his place of sleeping and napping: on the sofa, the armchairs, the chairs of the table on the porch, the table itself (among the plant pots on it), the kitchen chairs, and the kitchen window-sill. My impression is that he likes the sofa best (and there too he alters the place where he lies), but I don't think that there is any constraint in his making these choices.

Interruption of the flow of behavior

Sitting before leaping: When Aviva is sitting in her armchair and calls Max to her, usually he comes to her running. But almost every time he stops his run in front of her right foot and changes to the seated sphinx posture (forelegs stretched out and rear seated on the floor). After some time (about a minute), whether Aviva calls him again – "Come Max, come to me you beautiful cat" – or not, the cat will jump onto her knees (always from the side of the right foot). Such behavior recurs also when the general direction of his walk/run is toward the sofa. At the sofa's feet he stops short his flow of movement, and after about a minute he jumps onto it. (But here I must note that on several occasions I have seen Max jump onto the sofa, the armchairs, and the telephone stand in a direct continuous flow of movement.)

Do these cases, where the continuous movement is interrupted before the leap, attest to thinking, weighing up, prior to concluding the entire movement, prior to attaining the goal? I do not think that the running-halting-jumping is akin to the pauses the cat makes when on the hunt, because in these cases there is no prey that alters the speed of his running or that stops – a response that stimulates in the cat's chase a reconnaissance pause (see Taylor, 1986; Tabor, 1997; Leyhausen, 1979). Furthermore, I do not think that these behaviors are of the same kind as the chain of responses always produced (like a reflex or an instinct) by the same stimulus because interrupting the flow of behavior appears in various situations and at a partial frequency. From here I learn that it is Max who decides when to interrupt the behavior, with the (anthropomorphic) purpose of "evaluating the new situation".

Is there further evidence that Max grasps a dynamic situation, a chain of events, and from it learns what has to be done in order to achieve his goals? I believe that the answer is affirmative. All the following episodes, which I discussed above – **ambush for the night moth, ambush for Aviva, armchair-catch** and **petting** or **chasing games, the tail game, visit to the vet**, and **sleep: running-rolling over onto the back** – are dynamic situations in which an interaction is created between Max and ourselves (Aviva and me). In all these episodes Max perceived the happenings as a continuum of events in which event (b) followed in time and space event (a), and he learned how to exploit them to achieve his goal.

So as not to repeat myself ad infinitum, I shall examine the last episode **sleep: running-rolling over on the back**. How can this episode be interpreted? First I shall describe it again, very briefly. After I have finished watching television, have turned off the TV set, and am en route to my room to go to bed, Max suddenly passes me running, and rolls over onto his back in front of my feet, fixing me with his eyes. I immediately stoop and pet him for many minutes.

In general this behavior is hard to see as an instinctive expression, as it does not take place every time I get up from the armchair and walk toward my room. When the episode occurs (as stated, at a not high frequency) it takes place at night (between midnight and two a.m.) when Max and I are in the lounge (Aviva has gone to her room), Max is snoozing on the sofa, and I decide to terminate my staring at the television and go off to my room. Why does Max do this? There are several answers. He wants to be petted, he wants games, or he does not want to be on his own. I do not believe that Max wants games, because here his behavior is different from his games behavior – see the description above of the **armchair-catch** and **petting** or **chase games**. Nor do I believe that Max wants petting particularly, because in the evenings, as I described above (**jumping on knees** and **scratching the throat**), when he indeed wants to be petted he jumps onto my knees. It seems to me that Max wants company, that he does not want to be left on his own. He has learned that at night, the moment I have turned off the TV set I go to my room and close the door, and he is left alone for many hours. (Incidentally, Max has also learned this: the moment he hears the phone ring he jumps off the knees.) Therefore, the moment I turn off the television and walk to my room he responds: he runs in front of me and lies on his back. I do not stay to entertain him, simply because I am very tired. But as compensation I pet him for long moments.

Infringing prohibitions

Entering the bedrooms: Aviva and I prevent Max from entering our bedrooms because he leaves hairs. The ban is accomplished by closing the door and chiding ("Max, scram"). Max has learned, then, that he is forbidden to go into the bedrooms. Nevertheless, his curiosity compels him to visit our rooms when we forget to close the door. From time to time, when my bedroom door is left open and I am watching television, I see Max going off toward my room. As he walks I raise my voice and say, "Max, don't go in there". The cat wags his tail, sometimes sits on his rear, and sometimes retraces his steps and snuggles down again under the table on the porch. Sometimes he goes into Aviva's room and flees when he hears her steps approaching the room, or her chiding. Sometimes Max sits on the threshold of Aviva's room as she reads a book. But when she falls asleep the cat steals in and settles down, curled up on the chair next to her table. When she wakes up he runs out of the room.

At times, when I leave my room for the bathroom or toilet, Max steals into my room to scratch on the straw basket under the desk or to jump onto the table and smell the papers and the books. Hearing my approaching footsteps, before I re-enter the room Max slips out, making a kind of mewing sound that starts somewhat high and descends quickly. The interpretation I give to these cases is this: Max has learned and knows that he may not go into our bedrooms, but his curiosity leads him to visit them. The speed at which he slips out of the rooms, while making his mewing noise – on hearing footsteps drawing near – attests, in my view, to some feeling that we humans call 'guilt'. This interpretation accords somewhat with the perception of Allen (1997), that pinpointing errors is an important element in awareness of content. Following Dretske (1986), who described a cognitive-learning mechanism for mistaken representation, Allen suggests that responses such as surprise, confusion, and rapid learning attest that the individual has perceived that she has made a mistake. I am not certain that Max indeed feels 'guilt', but his responses may be subject to categorization as surprise, rapid learning, and confusion: he at once stops his sniffing around (like someone caught red-handed), and quits the forbidden room posthaste (like someone who knows what is expected of him), making a noise that sounds to me like atonement for his transgression.

Scratching the armchair and its legs: Although we fixed a post wrapped in a mat (mat-post) next to the toilet for Max to sharpen his claws on, and although we taught him to scratch only on the mat-post, and although today this post looks like something hit by a hurricane, with its strands hanging off it like worms, the cat has not left even a single armchair unscratched through and through. Max especially likes to scratch the upholstery of the seat of the armchair standing against the outer side of the wall of my bedroom. When Max did this in our presence we would scold him, and if he did not stop we would approach him with a threatening movement. The result was that he learned to do two things. First, he learned to scratch the armchair's wood leg instead of the upholstery itself (which was beginning to look like the mat-post), a change which, it turned out, led to a dramatic fall in our shouting at him; and second, he learned to scratch the chairs under the porch table – far from our field of vision and hearing. (When we do catch him doing it, we scold him and he stops scratching.)

In comparison with the foregoing case of **entering the bedrooms**, which may perhaps be interpreted as attesting to some component of fear-guilt, I think that in the present case Max simply learned to distinguish places that involve immediate sharp negative reactions from places that involve very few negative reactions. Still, I believe that this learning, based on giving negative reinforcement to a response, by its nature is linked to fear: it is reasonable to assume that the upholstery at the edge of the armchair seat arouses fear in him, and prevents him from scratching it more than scratching the armchair leg, to which, as stated, we hardly react.

The principle of new application

As may be seen from the account of the above episodes, (1) Max used existing (instinctive) behaviors, such as scratching, movement, posture, etc., to attain new goals. These goals concern interaction with us (Aviva and me) and may be summed up in very few words: Max is interested in our company and attention (petting); (2) Max uses various behaviors to achieve the same goal, for example, he wins petting from us by jumping onto our knees, onto the telephone stand, stretching his neck, rolling over onto his back, and suchlike behaviors.

Here the following comment is apt. So far, my observations and interpretations of Max's behavior have been made from my viewpoint. Now it is worth asking the opposite question: What is Max's response to our behavior (Aviva's and mine)? How does he take our behavior in his living space? These are very difficult questions, because the answers depend on the *interpretation we give to Max's behavior*.

The fact that on the one hand Max depends on us and has learned some of the above prohibitions, but on the other hand he treats the apartment as his own (e.g., he sleeps on the sofa, on the armchairs, or on the porch chairs – as he chooses), shows that a fairly complicated interaction of dominance exists between us: as may be seen from the foregoing, in many cases Max determines the start and finish of the interaction. Furthermore, as many of the above-described episodes show, Max has learned our habits and exploits them to attain his goals, e.g., being petted. Take the example of the episode **sleep: rolling over on the back**. If Max had not learned my habits (turning off the TV and going to bed), I do not think he would run in front of me and roll over on his back a second before I go off into my bedroom to sleep.

3.3 Discussion of indicators of free will

As I suggested above, *behavioral indicators* uphold the following conditions:

A. There is a theory T_p based on B_p, and empirical observations that support a connection of the kind $B_o = f(S, T_p)$, namely a family of connections exists between certain situations and certain responses that are explained by T_p (where B_p denotes private behavior, B_o denotes public behavior, and S denotes a situation of stimulus);

B. This connection is no other than a form of correlation, association, because S and B_o are likely to predict B_p well, but are not any kind of necessary and sufficient conditions for B_p.

Following this proposal, I will now posit general conditions answering the question when do we, human beings, treat a certain behavior as expressing free will. The answer is built on clear situations in which we are unwilling to ascribe free will to a given behavior:

Situation (a): The same individual, in the same state of stimulus, at different times responds with the same response.

Such cases, which include behaviors such as reflexes and instincts, are usually explained mechanistically (by an appeal to neurophysiological, genetic-evolutionary explanations). But if we obtain the following situation:

Situation (b): the same individual, in the same situation of stimulus, at different times responds with different responses,

we will have to seek other explanations because it is not possible that for the same individual in the same state of stimulus, a law or scientific theory will produce different predictions – responses or results. In this case, in which a theory predicts a wide range of responses of different kinds, no possibility exists of developing science and proposing a scientific explanation because in fact such a theory predicts all possibilities, and therefore it explains nothing.

In such cases we have to suggest that different responses have been obtained because in fact *the individual has changed.* We ascribe free will to her in the case where the state of stimulus remains fixed, but the individual herself has changed and her response has changed. But even in this case we must limit ourselves to a certain kind of variability in the individual. The change does not include neurophysiological changes and illnesses, which put the individual into situations in which she can only respond in one certain way. The development of an allergy, adolescence, aging, and processes of decay are instances showing that the same individual changes her behavior in an extreme manner at different times. (It is very hard to decide whether to include mental illnesses here, or severe disturbances in behavior, because it is not clear if they can be explained fully by appealing to a mechanistic explanation, and also because it is hard to apply this concept to the behavior of the cat.)

The change in the individual that I refer to, then, is connected to changes in private behavior, B_p, in normal situations, that is, changes in feeling, in wishes, in thought, in considerations. So I suggest that a behavior that attests to free will exists in the following conditions:

> Free Will (FW) conditions: ***The same individual whose private behavior has changed*** responds, in the same state of stimulus, at different times, with ***different responses.***

So far I have concentrated in the discussion on behavior through free will on four important components: the individual, the state of stimulus, the response, and the time. Now I shall introduce into the present analysis an additional component: the function or goal that the response realizes. This addition makes it possible to deal with the complexity that characterizes the everyday behavior of Max the cat, behavior that is beyond simple behavior, such as reflexes, which may be described as follows: the same response given by the same individual in the same state of stimulus at different times. (Here it is worth noting that it is also possible to assign to a simple reflex a certain adaptive-evolutionary function.) In this context I shall distinguish two behavioral conditions, which accord with the Principle of New Application (see above) and which attest to free will:

A. ***The same individual whose private behavior has changed*** responds, in the same state of stimulus, at different times, with different responses to achieve ***the same goal***;

B. ***The same individual whose private behavior has changed*** responds, in the same state of stimulus, at different times, with the same response to achieve ***different goals***.

As an example of the first condition let us look at David, in Haifa, who wants with all his might to reach Tel Aviv (for an important meeting) in any way possible: riding in a cab, on a motor scooter, on a motorbike, on the bus, or in a speedboat, or even by plane. In this case David tries to realize his objective by different responses. That is, David examines all the possibilities of achieving his goal according to considerations of efficiency.

As an example of the second condition, we shall look at David who has driven in his car from Haifa to Tel Aviv: once to meet the managing director of the Far-Talk mobile phone company, once to hear a Mozart concert, and once to get away from Uzzi, to whom he owes a large sum of money. In all these cases the stimulus-environmental state has not changed, so neither has the response – traveling in his car. What has changed, what explains David's journey each time, is his mental state, his private behavior. What moves David the first time is a business matter, the second time love of music, and the third time unpleasantness and fear.

In both these examples David acted out of his free will (even in the last case David could have decided differently, to stay in Haifa and settle his debts). Similarly, as arises from the discussion of the Principle of New Application – see above, it is possible to see that in several behavioral episodes the cat realized different goals by means of the same response, and the same goal by means of different responses.

So we can expand the FW conditions in the following way:

> ***The same individual whose private behavior has changed*** responds, in the same state of stimulus, at different times, with ***different responses*** to realize the ***same goal***, or with the ***same response*** to realize ***different goals***.

As we have no possibility of conducting direct observations of the private behavior of another (only I can observe my own mind, that is, engage in introspection), I suggest, in pursuance of the methodological criterion I suggested in the first chapter, namely equal hypotheses testing, that if FW conditions are realized, and if it is hard to suggest a reasonable mechanistic explanation for this situation, this behavior may be regarded as an expression of free will, and an explanation may be proposed based on appeal to private behavior of the same individual, that is, based on appeal to the explanations prevalent in everyday psychology.

Here it has to be re-emphasized that the status of the mechanistic or the mentalistic explanation is no more than an explanatory hypothesis which withstands the test of observation. For this reason, before suggesting a mentalistic explanation one must examine meticulously if a given episode is not susceptible to a mechanistic explanation. As some of the episodes I gave above as attesting to free will may well be interpreted as based on learning, so that mechanistic explanations can be suggested for them, I shall now move on to discuss these episodes.

Observational learning (imitation) and social learning in animals

In this section I shall concentrate on the same dynamic episodes which, in my opinion, Max perceived as a continuum of events in which event (b) follows in time and space event (a) (without attributing to Max a philosophical understanding of the notion of causality), and learned how to exploit them to achieve his goal: **ambush for the night moth, ambush for Aviva, armchair-catch** and **petting or chasing games, the tail game, visit to the vet**, and **sleep: running-rolling over onto the back**.

Can these episodes be explained by appeal to accepted mechanistic explanations, as with the explanation of observational learning? (Observational learning has a respected long-standing research record with animals and humans; see, e.g., Bandura, 1986; Hilgard & Bower, 1966; Mackintosh, 1974; Shettleworth, 1998.) For example, Meltzoff (1996) suggested that human infants are endowed with an innate urge to imitate humans; Boyd & Richerson (1988) suggested that imitation of behavior constitutes a learning mechanism which compensates on the one hand for inflexibility in genetically structured behavior in animals (birds), and on the other hand for the slow change in behavior as a result of trial-and-error learning; and Shettleworth (1998) considered the possibility that a special learning mechanism exists in social observational learning. (The literature also discusses whether this kind of observational learning (imitation) is *sui generis* or if it can be reduced to the typical explanations of classical learning (Pavlovian) and/or of instrumental learning (Skinnerian). This discussion exceeds the purposes of this book.)

To substantiate the importance of this learning for our concern, we shall look at the following experiments carried out with cats (see John, Chesler, Bartlett & Victor, 1968). The cats ("observers") watched other cats ("demonstrators") learning to avoid an electric shock by jumping over a hurdle. The observers learned this avoidance quicker than a control group that was not exposed to the demonstrators. Further, cats that watched the demonstrators learning to press a lever in order to obtain positive reinforcement (food) learned this pressing quicker than an appropriate control group (demonstrators that received reinforcement without pressing on a lever). So observational learning considers an observer who learns to imitate the action of the demonstrator, learning which includes, among other things, the sequence of the demonstrator's actions that yield the reinforcement.

Is it possible, for example, to regard the **armchair-catch game** as a kind of observational learning? If the answer is affirmative, in principle this episode may be explained by an appeal to one of the mechanistic explanations customarily proposed for this kind of learning. If the answer is negative it becomes more feasible to understand this episode from the viewpoint of free will and to suggest an explanation suitable for this episode. In my opinion, this behavior should not be seen as a kind of observational learning.

In most experiments in observational learning, the observer watches the behavior of her own species, and tries to copy it. In the episodes of Max, the cat watched our behavior. Is it possible that Max learned to imitate behavior of creatures not of his kind? I my opinion the answer is affirmative. Max, who has been in our company for about nine years, since he was a kitten, got used to us and became attached to us, and as a result he imitated some of our behavior. Two pieces of evidence support this answer. **Head on the cushion**: In a large number of instances I found that Max slept on his side with his body on the sofa and only his head on the cushion. **Speech-like mewing**: In many cases Max turned to us or responded to our talking to him (especially to Aviva) with mewing and making sound of various kinds and quality. Frequently we try to imitate his mewing and as a result a cat-like dialogue develops between us. In observations I made of alley cats I did not discern this form of communication (except, of course, for occasions of alley cats in heat.) Again, it seems to me that the cat is trying to imitate our verbal behavior: he has learned that communication between Aviva and me is verbal, and when he wants to communicate with us he addresses us with speech-like mewing. Despite support for the argument that Max is capable of learning from creatures not of his kind, I believe that it is hard to explain the episode of the **armchair-catch game** as observational learning.

Like learning, observational learning, imitation, is based on the fact that the observer has acquired new information and that her behavior has changed accordingly. In light of this, the question arises as to what in fact Max learned to imitate in the episode of the **armchair-catch game**. Did Max learn to approach the armchair and sit-lie in it? Does this learning explain this episode? To my mind the answer is negative.

Max has jumped into and sprawled on my armchair, on Aviva's, on the sofa, on the chairs on the porch, and on the table on the porch since the time he was brought into our apartment. I do not think that Max learned this from me (or from Aviva) by imitating, because these behaviors appeared while he was still a kitten, long before we played the **armchair-catch game**.

Furthermore, this game contains several behavioral elements that do not appear in his approaching the armchair and sitting-lying in it. (a) Max does not just go up to the armchair and sit-lie in it, as he did in many cases when I found him lying in my armchair or when I saw him jumping into Aviva's armchair. In the present case Max jumps into the armchair just an instant before I make to sit down in it. (b) In the present case Max does not sit or lie in the armchair (after sniffing at its seat), but stands on his four legs in a taut posture, intensely ready to leap from the armchair. (In the **armchair-petting game** Max at once rolls over on his back and expects me to stroke and tickle him.) This is a response that Max has not learned from us – Aviva and I simply don't play "musical chairs".

Then how may the **armchair-catch game** be explained? I believe that this explanation is based on two main stages. At the first stage Max learned the following chain of events as a single episode: *Sitting in the armchair*: Sam walks toward his chair and sits down in it (to read or watch television). Over the years, this order of events has occurred thousands of times before Max. I shall call learning of this kind "learning of environmental dynamics". This is perceptual learning, accomplished through the animal's watching changes taking place in its surroundings, whereby it learns the temporal and spatial connection of events in its environment.

Here it is worth noting that many animals are able to learn complex patterns of environmental stimuli and events, such as learning a serial pattern of stimuli, of what leads to what, of a spatial map, and to respond to these patterns accordingly. An explanation of these complex kinds of learning posits that this environmental dynamics is represented in the cognitive system of the animal, which uses this representation when it is active (see discussions in Domjan, 1998; Shettleworth, 1998).

In the second stage Max used this information, *sitting in the armchair*, to attain his goal – to play catch (or to be petted). To convey to me the message "I want to play", Max jumped a second before I sat down in the armchair and stood in it in his ready-to-run posture. (Often, as I described in the previous chapter, the game continued with a very enjoyable chase. This **armchair-catch game** is similar to the **chase game** described above.)

According to this interpretation, then, Max made an interesting connection between the following three elements: (a) his goal of playing with me, (b) the knowledge of *sitting in the armchair*, (c) attracting my attention by jumping into the armchair and assuming a ready-to-run pose.

Are there more episodes that may be interpreted as learning of environmental dynamics? The answer is affirmative. In a number of episodes that I described above, for example, **ambush for the night moth, ambush for Aviva, visit to the vet**, and **sleep: running-rolling onto the back**, Max used information about the different happenings that took place in his environment (the flight of the moths around the night lamp, Aviva's movements in the apartment, the series of events connected with the visit to the vet, and my habits before I go to bed) to win attention, petting, and other social goals.

To conclude this matter I will say that the **armchair-catch game** is not a kind of imitation behavior. Although an important element in the explanation of this behavior is connected with observation and learning my habits of behavior (observation that appears in other imitation learning too), other elements are linked to Max's will and goals, and to the successful path that led him to realize his intentions.

3.4 Indicators of free will, consciousness, and explanation

In this section I shall propose two ideas: (a) that free will behavior in Max attests that this behavior is performed consciously, and (b) that this attribution of consciousness to the cat makes it possible to explain some of his behavior by appeal to consciousness concepts such as intention, will, and knowledge. In other words, if we feel that just as we appeal to our inner world to explain a considerable part of our behavior we should do the same to explain the cat's behavior (i.e., refer to his inner world), we are bound to vindicate the argument that the cat, like us, behaves out of awareness, consciousness.

Assuming that the above behavioral indicators are acceptable as expressing free will in Max, it is possible to learn that part of Max's behavior is performed out of consciousness. Support for this idea is based on the analogous argument called *Free Will – Consciousness*. First I shall present the argument briefly, and then I shall discuss it in detail.

1. *In humans*, according to the phenomenological-subjective approach, free will behavior is performed consciously.
2. *In humans*, behaviors to which free will is ascribed according to the behavioral approach, that is, according to the FW conditions, are also behaviors to which free will is ascribed according to the phenomenological-subjective approach.
3. *In the cat*, the behaviors that meet the requirements of FW conditions constitute indicators of free will.
4. *In the cat*, it is possible to suggest on the basis of 1, 2, and 3 that these behaviors are performed consciously (as understood by us, humans).

First I shall discuss the first two parts of the argument. The behavior of human beings may be divided into two basic categories: conscious behavior (that which is conscious, or that which is likely to be conscious), and unconscious behavior (that which will never be conscious) (e.g., Rakover, 1983a, 1996). Only some of conscious behavior may be regarded as free will. That is, not all conscious behavior is necessarily free will behavior. For example, I am aware that I have made a knee reflex response and that it happened despite my will. I am aware that in the Müller-Lyer illusion <>—< the left horizontal line is perceived as shorter than the right horizontal line, even though I am aware that these two lines are of equal length. I am aware that I am imitating a person who talks in Cockney English even though I am not aware of the action of the (motor-nerve) system responsible for producing this form of speech. In my army basic training I was aware that I was performing the most body-racking exercises contrary to my will and often without my wishing to be involved in it at all.

It is hard to imagine that an act of free will may be done without the individual being aware of an important part of this behavior, without awareness, consciousness, not being involved in this act. For example, I am aware that I intend to write this sentence, and as I write it I am aware of the movement of my fingers striking the computer keyboard, of the text that appears before me on the monitor and its meaning, and so on. Yet I am not aware of the neurophysiological processes connected with the movements of the fingers and their link with the eyes and brain, and all the complicated cognitive activity going on in my brain and that is responsible for this sentence being written according to the rules of language and grammar. Libet (2002) in his experiments defines free will operationally in keeping with the prevalent opinion:

> First, there should be no external control or cues to affect the occurrence or emergence of the voluntary act under study; that is, it should be endogenous. Second, the subject should feel that he or she wanted to do it, on her or his own initiative, and feel he or she could control what is being done, when to do it or not to do it. (p. 552)

From this definition it clearly emerges that awareness of acts is a vital part of free will because otherwise it would not be possible, according to the prevalent view, to understand such notions as "voluntary act"; "should feel"; "wanted to do"; "own initiative"; "could control". These phrases, as is readily seen, refer to the individual's introspective world (to the individual's reporting her observing her internal world): to her awareness of her wishes, her feelings, and her beliefs.

Does the behaviorist perception, which does not rely on this introspection (because according to behaviorism introspection does not meet the scientific requirements of observation: see Rakover, 1990) accord with the perception of free will from the phenomenological-subjective viewpoint of the individual herself? To examine this matter I shall again present the behaviorist criterion of free will that I developed above:

FW conditions: ***The same individual, whose private behavior has changed***, in the same state of stimulus, responds at different times with ***different responses*** to realize the ***same goal***, or with the ***same response*** to realize ***different goals***.

The change in the individual I refer to is connected to changes in normal private behavior, that is, changes in feeling, will, thought, and consideration.

I hold that if the same individual, at different times, in the same state of stimulus, does *not* respond with different responses or does *not* realize different goals, then one cannot speak of free will behavior because this behavior does not accord with Libet's definition: "…the subject should feel that he or she wanted to do it, on her or his own initiative, and feel he or she could control what is being done, when to do it or not to do it". Nor do we obtain a change of behavior as an expression of free will when in the same state of stimulus neurophysiological changes happen to the individual (e.g., illnesses) that are subject to scientific observation, and that constrain her to behave in this way and not any other. This is because this situation too does not accord with what Libet suggests: "… there should be no external control or cues to affect the occurrence or emergence of the voluntary act under study; that is, it should be endogenous".

The upshot, then, is that attributing the concept of free will to a certain behavior according to FW conditions does not run counter to attributing this concept to the same behavior according to the phenomenological-subjective perception.

Still, it may be argued that FW conditions are too broad and allow the inclusion of behavior that does not clearly attest to free will. We may look at an advanced robot, which chooses alternative (a) over alternative (b) in the morning and reverses its choice at night. For the onlooker this behavior seems like an expression of free will. (This example is no more than a kind of variation on the Turing test, on which I shall expand in the following chapters.) I do not believe it is. A careful check of the specifications of the programs and software by which the robot was assembled will show that a mechanistic explanation can be suggested to separate the day/night behavior of our robot. Not so with human beings: after we ascertain that the state of stimulus is the same, that no neurophysiological changes have taken place that impose on the individual an uncontrollable behavior, the change in the individual's behavior will be explained by resorting to her private, conscious behavior, that is, to free will.

Now I move on to consider the two last parts of the argument. Human behavior, which is performed out of consciousness and to which the property of free will is ascribed according to the phenomenological-subjective approach, also satisfies the free will requirements according to FW conditions. Since the behavioral indicators attest that Max's behavior was performed out of free will according to the FW conditions, it may be proposed analogously that the free will behavior in Max was performed consciously. In other words, the core of the analogy argument is constructed from the following elements:

In humans: Free will behaviors are connected to consciousness.

In Max: Certain behaviors attest to free will.

Analogy between Max and humans: These certain behaviors are conscious.

If this analogy argument holds, and some of Max's behavior is performed through consciousness (when his consciousness is understood as possessing similar properties to human consciousness that accompanies free will), we are now able to go on to the next step and suggest that his behavior can be explained by appeal to the concept of consciousness as an explanatory concept. That is, we can now suggest that part of Max's behavior will be explained by addressing his intentions, his desires, his goals, and the knowledge that he has acquired throughout his life. (On the distinction between consciousness as a phenomenon requiring an explanation and consciousness as an explanatory concept, see below.) (Here it is worth stressing that because this analogy argument – free will-consciousness – is not binding logically, this analogy is to be treated as a hypothesis for an empirical test.)

Kinds and levels of consciousness

Is it possible to find other evidence, other arguments, supporting the hypothesis that animals (including Max the cat) are aware, are conscious, of their behavior and of their environment? In my opinion the answer is affirmative, but as this is a question whose answer is complicated and calls for a review and discussion of a large number of observations and experiments with many animals, an effort that exceeds the purposes of this book, I shall comment only briefly as follows. (On this matter see Allen & Bekoff, 1997; Griffin, 1981, 2001; Levy & Levy, 2002.)

I maintain that consciousness (on the minimal level – see below) is a necessary and sufficient condition for discerning between living creatures and other organic and inorganic matter or objects based on dynamic systems such as cars and computers. Without consciousness of pain (a feeling of pain) animals would not respond as they do, and their behavior would not be in any way different from a broken wooden leg of a table, a computer that has fallen to the floor and shattered, or a car that has crashed into a wall. Without consciousness (in its different variations and degrees) our relation with the world around us would be exactly the same as that of a pile of earth to the field around it and to the ants burrowing in it. From the viewpoint of this pile of earth (so to speak) there would be no significance in its position or in the chain of events in which it is located. From this standpoint I hold that no living creature exists that is not endowed with some degree or other of the most marvelous quality of all – being aware of the world, being in a state of consciousness of the world.

This description may be viewed as a kind of a 'thought experiment', which suggests that unconscious behavior is no more than a group of physical, chemical, and neurophysiological processes – a proposal that requires some further clarification. LeDoux (1996), who discusses in depth fear behavior in his book *The emotional brain*, suggests that fear responses such as freezing, escape, fight, facial expressions, hair bristling, changes in blood pressure and pulse rate, can appear without awareness; that is, without the involvement of the prefrontal cortex, which is assumed to be connected with consciousness generation, in the production of fear behavior. These responses, which are considered universal since they appear in different species of animals, can be elicited directly and unconsciously by another brain area called the amygdale. (LeDoux also reviews rich literature that supports the hypothesis that a large part of the emotional and cognitive processes occur without awareness.) I have no dispute with this, and have already mentioned (see chapter 1) that consciousness evolved on the basis of a very complex neurophysiological and unconscious activity. What I would like to stress here once again is that without awareness behavior is no more than chemical and physiological reactions; and that animals possess different degrees of consciousness, which are connected with different levels of information processing of stimuli, contexts, responses, response feedbacks, and the appropriate activity of the neurophysiological systems.

LeDoux also suggests that the experience of fear is generated when the outputs of the neurophysiological system, which is involved in the production of fear responses, are represented in the system generating consciousness, that is, in the system known by the name 'short-term memory' or 'working memory' – the cognitive system associated with consciousness. I have no problem with this proposal either, and basically there is no significant difference between LeDoux's position and mine.

The difference lies, I think, in LeDoux's over-emphasizing the importance of the body and brain emotional systems, so far as to over-minimizing the importance of consciousness: "The states and bodily responses are the fundamental facts of an emotion, and the conscious feelings are the frills that have added icing the emotional cake" (p. 302). I disagree with this approach. I don't have any doubts about the crucial importance of the neurophysiological system for consciousness. But without consciousness I would not feel and know that I am writing this sentence. I would be like a plant writing poetry by shedding leaves. By comparison, the methodological dualism developed here places the two processes, conscious and unconscious, on the same epistemological level and proposes that to understand an animal's complex behavior one has to employ two schemes of explanations: mechanistic and mentalistic.

From these viewpoints then, the question is not if Max is endowed with consciousness, but with what kind of consciousness he is endowed, and to what degree.

I think that Max is aware of the information that his senses supply him with (consciousness similar to what Griffin, 2001 terms 'perceptual awareness'). But it is hard to assume that he has self-consciousness, that he is aware of his awareness. Such a high degree of reflective awareness exists in humans and possibly also in monkeys and dolphins. For example, Smith, Shields & Washburn (2003) suggest on the basis of a series of experiments that like humans, monkeys and dolphins exhibit conscious processes of self-supervision and self-adjustment. In these experiments, which indicate cognitive control of learning processes, all the subjects reported, by performing a special response, their difficulty in learning complex distinctions in a similar manner. These researchers suggest that conscious processes go into action in situations where the individual is required to rework difficult and elaborate information. As supporting evidence of this notion, they cite William James, stating that consciousness is aroused when the individual is about to perform a dangerous act (e.g., a cat's risky leaping). (For the interested reader the paper by Smith, Shields & Washburn, 2003 is a target article and has appended to it responses, criticism, and the authors' reply.)

Bekoff (2002) writes:

> If 'being conscious' means only that one is aware of one's surroundings, then many animals are obviously conscious. Simple awareness of this sort is called 'perceptual consciousness'. (p. 92)

In Bekoff's view the interesting question is if animals are bestowed with the ability to possess self-consciousness (and see Griffin, 2001). Max may not be endowed with this high capacity for self-consciousness and consciousness of consciousness, because these consciousnesses depend on the development of language. Still, it may be suggested that Max has additional consciousnesses beyond sensory-perceptual awareness, and that he may be conscious of his desires.

Body awareness. It is hard to believe that the cat is not aware of his body, which he licks incessantly (even thought this behavior is perceived as innate, characteristic of cats); that he is not aware of sharpening his claws on the furniture and honing them on his teeth. Moreover, I assume that Max is aware of the harm his claws are liable to cause – otherwise it would be hard to understand why he is careful not to extend them when he plays with us. Max doesn't like his paws to be touched, but he lets us clean up his eyes willingly. (Sometimes Max tends to purr when we clean his eyes. So this may be seen as a further example of the Principle of New Application; purring, which appears at the time of suckling, now appears when his eyes are attended to and also when he is stroked.) Max, then, treats the parts of his body in different ways. Finally, Max responds when his name is spoken.

Consciousness of another. It is hard to support the hypothesis that Max has a 'theory of mind' on consciousness of the other (e.g., the consciousnesses of Aviva and of me), a theory by which he explains another's behavior. This is very hard to test empirically. For example, in the episode of **sleep: running-rolling over onto the back**

Max acts in keeping with my behavior to achieve his goals. Does Max act according to the 'theory of mind' (Max thinks: Here's this Sam, tired and wanting to go to bed, so he turned off the TV)? Or does he act through learning a series of stimuli, a series of events, the *episode of Sam's bedtime*, which he utilizes to attain his goal: not to be left alone, to get petted? Evidently, for now there is no answer to this question.

In a review and discussion of the subject of 'levels of consciousness', Piggins & Phillips (1998) suggest a four-level scale: sensation, sensation/perception, perception/cognition, cognition; these are ranked according to the level of activity of the central nervous system, complexity of the behavioral phenomenon, mental processes, and processes of reworking information necessary to deal with the given phenomenon. The demarcations in this scale, as Piggins & Phillips admit, are not clear and incisive. Furthermore, the concept of complexity is not fixed and depends on our level of knowledge; for example, the notions of atom and cell were once taken as simple foundation stones for understanding more complicated phenomena, while today we know that these two terms are nothing more than titles for a rich world, whose measure of complexity is enormous.

Similarly to what has been stated above, they assert:

> Our view is that all animals possess some consciousness, and its correlate awareness, however crude, or transient this may be in relation to man and however inadequate our use of consciousness and awareness as synonyms is to some. It is presumed that whatever degree of awareness other creatures possess, it is appropriate to their needs. (p. 186)

It is reasonable, therefore, to assume that the very existence of consciousness (in its degrees) as a condition for the essence of a live creature explains the great difficulty in attempting to operate consciousness as an independent variable – for example, to compare conscious behavior with unconscious behavior. On this matter it is worth noting that people who lose consciousness for a lengthy time (due to accident or stroke) and are kept alive artificially, are perceived as vegetables.

The problem of 'the other's mind (consciousness)'. Griffin (2001) writes:

> Many philosophers have wrestled with the question of how we can know anything about the minds of others, whether they be other people, animals, extraterrestrial creatures, or artifacts such as computers. (p. 255)

This 'problem of the other's mind' has sparked a wide-ranging philosophical debate, which I shall not discuss here extensively. The problem is rooted in the philosophical approach of Descartes, holding that we do not experience-know anything except our inner world, and we cannot experience-know the private world of another; therefore, we are unable to justify the belief that another has a mind and consciousness (see, e.g., the discussion by Pinchin 1990). Furthermore, the inner world is that which makes each and every one of us possessor of a unique personality. Deprive David of his mind and what is left is a body deprived of uniqueness subject to public observation.

One of the solutions to this problem is given by the argument from analogy, which as we shall see later is an important component in our treatment of animals:

I look at myself: in state of stimulus, S, an inner response is aroused in me, a mental state, MS, which arouses a public response, R_p, that is, S – (MS) –R_p;

I look at another: S – (X) –R_p, that is, I observe S and R_p, but I am not able to look at the inner response, at the mental state, of the other;

Analogical inference: As the state of stimulus and the public response in me and in the other are similar, I infer that X = MS. (This inference offers a solution to two problems at once: it suggests that the other has a consciousness, whose content is similar to that of my consciousness.)

This inference has generated wide philosophical and empirical criticism (e.g., Pinchin, 1990; Provinelli & Giambrone, 1999). First, the analogical inference is not as certain as the deductive inference. Secondly, doubts have been voiced as to the causal connection S – (MS) – R_p: this connection is not by way of a natural law; and as I described in chapter 1, some cast doubt on the causal relationship between the mental state and the internal response. Thirdly, there is a danger of a serious error of generalization of one case (I, myself) to others. And fourthly, Provinelli & Giambrone have shown in a series of experiments that there is no need to ascribe high mental processes to chimpanzees to explain their behavior, since better explanations can be suggested based on simple processes. On the grounds of this set of experiments these authors determined that chimps do not have a 'theory of mind' as we humans do, that is, a theory whereby the monkey explains the behavior of another by addressing the other's mental states, for example, her feelings, her desires, her thoughts, and her intentions.

To these criticisms and others the following point may be added. The analogical inference, which is based on finding a similar feature in responses of the human and the animal, is inappropriate for the modus operandi of science. For example, while it is hard to find what the similar feature (visual, categorical, etc.) is for the movement of the moon around Earth, the fall of a book from a shelf, oscillation, pushing a wardrobe full of clothes, and high-jumping, it transpired that all these phenomena received a full explanation in the framework of Newton's theory.

From this angle it seems, as I suggested above, that the analogous inference should be taken as a research hypothesis subject to confirmation and refutation. For example, the explanation for the escape behavior of Max the cat and of escape behavior in a similar state of stimulus (a dog barking) may be based on a similar mental state – the experience of fear. In other words, the above inference X = MS, whereby the other is endowed with MS similar to mine, is no more than a kind of hypothesis that explains both my behavior and the behavior of others.

This view of the analogous inference is linked to a known type of inference that seeks the best explanation for a given phenomenon (this is called 'inference to the best explanation', which I mentioned in chapter 2: see Josephson & Josephson, 1994; Lipton, 1991, 2001a, 2001b. Note that the philosophical literature finds interesting connections between analogy, induction, and inference to the best explanation). From the viewpoint of this inference, the analogy argument suggests that the best explanation is the hypothesis that the other too is endowed with a mental state similar to mine. But it should be stressed again that this is no more than a suggestion, and it is very likely that hypotheses will be found whose explanatory power is greater than the analogy hypothesis (e.g., as suggested by Provinelli & Giambrone, 1999).

Explanation and consciousness

The general impression is that learning in humans is done through awareness (e.g., learning at school, at university, learning the rules of driving, etc.), while reflexive and instinctive responses are unconnected to awareness. Does it follow that learning is to be explained by appeal to a mentalistic explanation, based essentially on the individual's conscious intentions, goals and knowledge?

Griffin (2001) writes:

> It is commonly assumed that conscious mental states can only be based on learning and that behavior that has arisen through evolutionary selection cannot entail consciousness. (p. 278)
>
> The assumption that only learned behavior can be accompanied by conscious thinking arises, I suspect, from analogies to our own situation. (p. 279)

If it is found that an animal's behavior is not subject to a genetic-evolutionary explanation, it will be possible to suggest a learning explanation connected, like that in the human, to consciousness.

This distinction is not clear-cut. First, a number of experiments on humans, called 'learning without awareness' or 'implicit learning', produced results that can be interpreted as learning without consciousness (e.g., Frensch & Runger, 2003; Rakover, 1993; Reber, 1993). Conversely, several experiments showed that training and skill turn responses learned out of awareness into automatic responses without planning and awareness (e.g., Shiffrin & Schneider, 1977). Furthermore, most explanations given to learning phenomena are mechanistic, and do not utilize concepts linked to the individual's internal mental world. (Even a "mental" theory, such as Tolman's, which uses concepts such as 'a cognitive map', are subject to an objective, not a subjective-mental, interpretation, by means of the theoretical notion of 'intervening variable'. See discussion in Hilgard & Bower, 1966.) In addition, Shettleworth (1998) argues that an important part of learned behavior is based on innate behavioral systems. From this aspect, it is clear that no behavior is either entirely learned or entirely innate.

Secondly, the individual is aware of some of the genetically programmed responses, even though they are performed automatically without the individual having control over them. As stated, it is hard to believe that Max is not aware of washing his fur, even though this response is part of the characteristic heredity of cats.

In light of this discussion, the requirement arises for methodological care in answering the question: what is the relation between consciousness and kind of explanation (mechanistic, mentalistic)? This relation, it emerges, is highly complex. Some of the reasons for this, it seems to me, lie in the following three factors.

A. Behavior is composed of a network of different responses, some that the individual is aware of, and some that the individual has no possibility of being aware of (see discussion above). The question of how such complex behavior may be broken down into parts subject to/not subject to awareness is a matter that will occupy us in the following chapters.

B. Consciousness as a behavioral phenomenon may be put to an explanatory discussion in three different ways. By the first, one refers to consciousness from the viewpoint of functionality. For example, one of the important functions of consciousness is the ability to represent reality, and by virtue of this to decide what is worth doing without taking any real risk.

The second way, the main way of explaining consciousness, followed by philosophy for many years, is to find a connection between body and soul, brain and mind, and to try to understand the mind in material terms (I shall discuss this issue extensively later).

The third way, an additional way for which I have not yet found systematic development in the literature, is to depart from the idea of levels of consciousness and try to define some "basic atom of consciousness" (sensory?) by means of which, through the development of suitable functions, it will be possible to explain complex consciousnesses. (Will it ultimately be possible to find a connection between this basic atom of consciousness and neurophysiological processes in the brain? I have my doubts.) Griffin (2001) suggests that the level of consciousness grows as a function of the degree of complexity of the nervous system.

C. Consciousness as a causal factor of behavior requires a distinction between two kinds of "explanatory consciousness": consciousness (a) and consciousness (b). By consciousness (a) I mean that the individual is aware of behavior over which she has no control and that she did not cause out of free will, like our awareness of the knee reflex or of a sneeze. Is this awareness an epiphenomenon (i.e., an event that has a cause but itself is not the cause of another event)? I do not believe so. Because of consciousness, we are capable of controlling and tracing our behavior and preparing for its results. For example, we are likely to apologize to the doctor when our leg kicks his arm, or for the sudden noise of the sneeze. In these cases consciousness was indeed not the cause of the response (knee reflex or sneeze) but it was the reason for the apology. Furthermore, in many cases awareness of instinctive behavior will probably influence or change it. For example, hunting behavior in the cat changes with great flexibility according to its perception of the dynamic change (sometimes unexpected) in the environment and in the behavior of the prey. It is difficult to imagine that such great flexibility comes about without consciousness, that is, that this flexibility is programmed in the cat genetically. It is therefore reasonable to assume that the consciousness itself is a result of evolutionary development and that its function, among other things, is to cope with rapid and unexpected environmental changes. Such flexibility is made possible through consciousness and not through genetic tailoring.

By consciousness (b) I mean that the individual is aware of behavior over which she has control and which she has initiated, caused out of free will; that is, consciousness is the reason for the behavior, for example, in the case of the apology, or in cases where the individual has realized her intentions and goals.

Finally, it is worth stressing that this interpretation of consciousness as an explanatory concept does not accord with other approaches. For example, Libet (2002) suggested on the basis of series of experiments that the main function of the mind is not to *initiate* behavior but to *supervise* it. Furthermore, several researchers maintain that the notion of consciousness in animals is superfluous as an alternative hypothesis can be raised, which is not based on consciousness, and which will explain mechanistically everything that is explained by the hypotheses based on consciousness. What is this alternative hypothesis? In an article on the subject entitled "Making room for consciousness", Radner & Radner (1989) answer:

> The usual answer is that cognitive psychology provides models of mental processes that make no appeal to consciousness. …When animal consciousness is dismissed as superfluous, we must ask whether the dismissal refers to consciousness as a phenomenon to be explained or as an explanatory device. The most plausible answer is that consciousness is superfluous in the latter role. Anything that can be explained by it can be explained equally well without it (p. 206). [I discuss this issue and its implications later on. See especially chapters 8 and 9.]

CHAPTER 4

The structure of mentalistic theory and the reasons for its use

The aim of this chapter is to resolve the question of why the methodological status of a mentalistic explanation is deemed lower than that of a mechanistic explanation, even though all the hypotheses are equivalent before the empirical test. The answer lies in a comparison between the structure of mechanistic theory and the structure of mentalistic theory: weaknesses were detected in the internal consistency of the latter, and in the connection between it and the observations. In that case, why should we use a mentalistic explanation? The answer indicated that there is a group of behaviors that rejects an exclusive mechanistic explanation. This group meets the criterion of mentalistic behavior, that is, behavior that requires a mentalistic explanation and does not accord with the criterion of mechanistic behavior.

No one published my new and revolutionary theory "Mind/body unity", so I had no alternative but to publish one part in *Voice of the Soul* and another part in *Sumo*.

In the previous chapters I proposed solving the problem of anthropomorphism by means of 'equal hypotheses testing'. The guideline according to this procedure is that mechanistic and mentalistic hypotheses are to be compared empirically, when their methodological status before the scientific test is equal. Despite this, the reader may point out that in many cases I proposed another guideline whereby a mentalistic hypothesis is to be used when it is hard to suggest a mechanistic explanation for the given behavior. Here then, the question arises of how these two guidelines conform to each other: if the methodological statuses of the hypotheses are equal, what point is there in the guideline that a mentalistic explanation is to be put forward only after a mechanistic explanation has failed? The answer is that this guideline does not refer to the stage of the empirical test itself. Now I shall expand this idea. The empirical treatment of a given hypothesis is based on three stages that I shall call *the three stages of the empirical test.*

At the *preparation* stage the researcher grounds the given hypothesis theoretically and empirically by emphasizing the connection of the hypothesis to existing theoretical and empirical knowledge. Only in this way can one show what is the contribution of the hypothesis to existing knowledge – by making a new prediction or by providing a new explanation for incomprehensible findings. In other words, scientists do not tend to take seriously "maverick" hypotheses devoid of any anchor to the existing body of science, or "disguised" hypotheses which prove to be no more than old and faulty ideas, and sometimes ideas that cannot be put to an empirical test.

At the *test* stage the researcher derives (a) prediction(s) from the hypothesis and checks whether the observation matches the prediction. The result of the test stage thus lies in the predicted-observed (p-o) gap, where a small gap attests to the soundness of the hypothesis.

At the *decision* stage the researcher examines the results of the test ((p-o) gap) taking into account the relevant theoretical-empirical knowledge, and reaches a broad and considered opinion on whether to reject-accept the hypothesis given the current state of knowledge (on rejection-acceptance of hypotheses see the discussion in Rakover, 1990).

Given these three stages, I argue that only at the second, the test stage, are the empirical hypotheses at the same methodological level, because at this stage the hypothesis is tested only by one criterion: the size of the (p-o) gap. This gap, as may be seen, is entirely indifferent to the scientific "pedigree" of the hypothesis, and tests it only according to the one single measure: its success in prediction.

The mechanistic hypotheses are thus ranked against the mentalistic at the two remaining stages, the preparation stage and the decision stage. At these stages broad considerations come into play (beyond the (p-o) gap), by means of which scientists treat the hypothesis, theory, from various viewpoints: the structure of a theory and its coherence, the measurements of theoretical concepts, and its integration in the theoretical-empirical background. These are also the stages at which profound scientific and philosophical views (including ethical and moral) influence the scientific evaluation of the hypothesis under consideration. For example, a researcher who has a deep religious belief that God created the human in His image and form will be very hard pressed to accept the theory of evolution and to believe that a cat acts out of consciousness and free will, like a person. (Note that Darwin, 1871/1982, held that the difference between human and animal consciousness is a matter of degree only.)

This proposal of three stages of the empirical test brings to mind somewhat the distinction drawn by philosophers of science between two kinds of scientific context: the context of discovery and the context of justification. While the former refers to discovery of new phenomena, to processes of creation of hypotheses and theories that explain a given phenomenon, the latter refers to processes of testing hypotheses – to methodology dealing with confirmation and refutation of hypotheses and theories (e.g., Losee, 1993). This distinction set off a debate and criticism, and today it is not accepted as clear-cut (see discussion in Hoyningen-Huene, 2006). Philosophy of science focuses on an attempt to understand not only the logic that underpins testing hypotheses but also the way in which scientists solve problems and develop explanations as part of the wide process of the advance of science (e.g., Bechtel & Richardson, 1993). So the second stage of my approach may be seen as the context of justification, and the first and third stages as conforming with the latter approach, namely with the attempt to understand the context of discovery as a highly important part in the development of science.

In light of all this, why have I proposed to rank a mentalistic hypothesis in second place and use it after it transpires that a mechanistic hypothesis fails to explain the phenomenon in question? The answer is connected to the scientific structure of a mechanistic, as distinct from a mentalistic, theory. The latter has several flaws, stemming from the fact that this theory does not meet scientific criteria. These flaws, as we shall see, are not fatal.

4.1 The structure of a theory

In this section I shall be helped by the following books and articles, which deal with the structure of a scientific theory: Bird (2000); Carver (2002); Rakover (1990); Rosenberg (2000); Suppe (1977). I cannot discuss all the subjects connected with scientific theory, so I shall concentrate on the aspects that are important for our concern. I suggest seeing an empirical theory as based on two levels, layers, of analysis: the theoretical and the observational.

The *observational* is a conceptual level that concerns observation and measurement of certain properties connected to the studied phenomenon (the cat's behavior); for example, the kind of stimulating environment to which the cat is exposed (the intensity of the visual, auditory stimulus), dilation and narrowing of the pupils, movement of the ears and tail, its motion (walking, jumping, rolling), and the speed of its reaction.

The *theoretical level* is likewise conceptual, representing the observational level and suggesting certain connections (e.g., logical) between its concepts. For example, a threatening-frightening stimulus arouses in the cat fear-attack reactions such as flattening of the ears, arching of the back, and crouching (e.g., Morris, 1997). I call this approach the *two-layer structure of a theory*.

Figure 4.1 depicts relations within the levels and between the levels. On the theoretical level we try to predict a behavior (r) by means of a process (p) that processes the stimulus (s) in a certain form, that is, r=f(s,p). The theoretical level, then, strives to reflect those processes, unknown to us, that occur in the individual (O) and that are responsible for her response (R) in a given state of stimuli (S). The observational level does not provide us with an answer to the question of what the process is that caused the individual to respond in a certain way to a given state of stimuli; but the theoretical level tries to offer precise details of this process. The theoretical level, in fact, offers a hypothesis stating that the process responsible for the individual's observed response may be like that detailed on the theoretical level. (Here I have to make an important comment. This formulation of the theory as based on two levels of analysis – observational and theoretical – does not entail total commitment to the 'received view' of the theoretical structure, which is based among other things on the distinction between the theoretical and observational concepts, or to any of a number of theoretical developments proposed as a solution to problems raised against the 'received view'. Furthermore, in this formulation there is no kind of exclusive commitment to the instrumental approach, that theories are nothing but tools for predicting phenomena in given conditions, but there is a certain tendency to agree with the realistic approach, that theoretical concepts do indeed represent entities, processes, that exist in nature, such as microbes and atoms. The reason for this lies in the assumption that the theoretical level seeks to reflect processes on the observational level. On these matters and others see reviews and discussions in the literature cited above.)

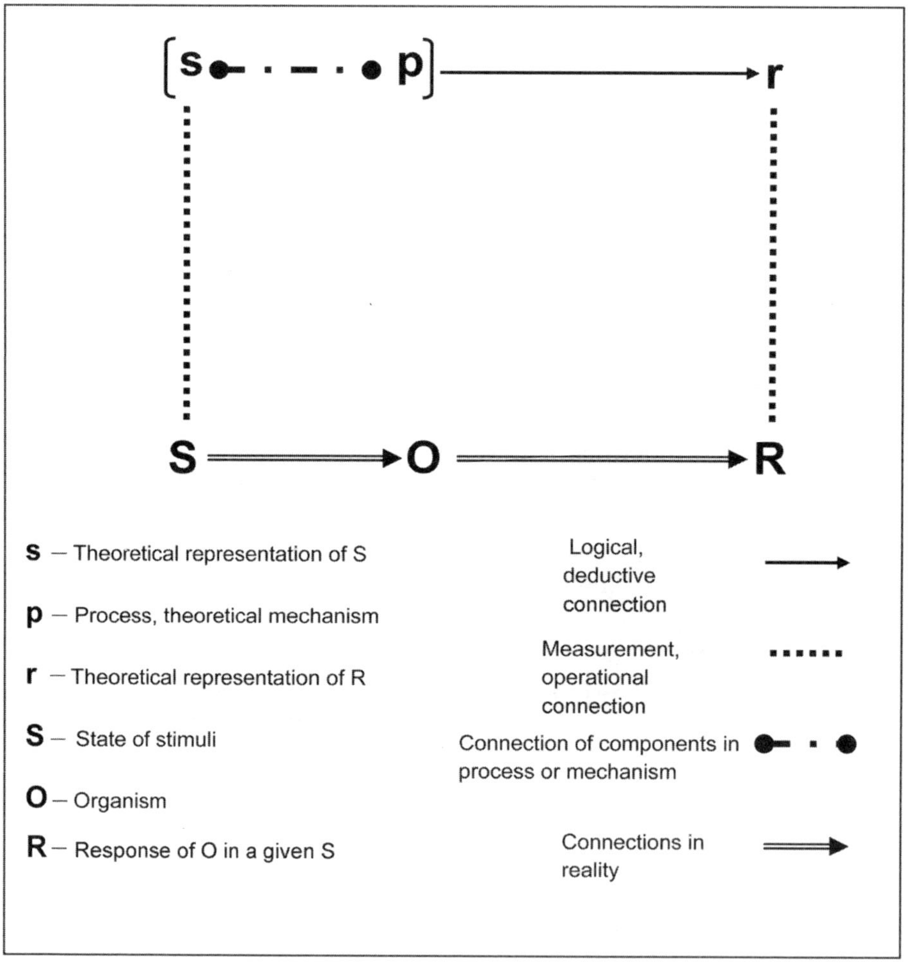

Figure 4.1. Diagram of a two-layer structure of a theory

In light of this perception of the structure of a scientific theory – the two-layer structure of a theory, I shall now attempt an answer to the question of the relation between the mechanistic hypothesis and the mentalistic hypothesis. I shall propose two criteria: consistency and measurement.

The first criterion (consistency) treats the question of consistency on the theoretical level. That is, to what extent are the terms on this level well defined and not open to different interpretations, and to what extent are the relations between the terms on this level defined unequivocally, so that given a certain stimulus the theoretical process will yield one response only. For simplicity I shall sort theories into two levels of consistency: complete consistency and incomplete consistency.

The second criterion (measurement) treats the connection between the terms (stimulus, response) on the theoretical level and those (stimulus, response) on the observational level. For simplicity I shall sort the theories into two degrees of measurement situations: valid and reliable measurement, and invalid and unreliable measurement. (The concepts of validity and reliability are defined in every introductory textbook to methodology in the social sciences. In brief, validity means that the measurement measures empirically only what is implied from the concept in the theory; reliability means that the measuring process measures the same thing at different times. See, e.g., Neal & Liebert, 1986.)

In consequence of these distinctions I propose three kinds of theory:

1. *Theories that meet the criteria of consistency and measurement*: These are mechanistic theories characteristic of the natural sciences.
2. *Theories that meet the criterion of consistency but not of measurement*: These are mechanistic theories formulated mathematically or by means of computer language, and are characteristic of experimental psychology and of behavioral models.
3. *Theories that do not meet the criteria of consistency and measurement*: These are mentalistic theories and are characteristic of what is called 'soft psychology' and everyday psychology.

My principal argument is this: mentalistic hypotheses, theories, are scientifically flawed because they do not meet either the consistency or the measurement criterion. As a result, when asked to explain a given behavior (of Max the cat) one first must check very carefully if it is not possible to explain this behavior by means of a mechanistic hypothesis. Only when the mechanistic hypothesis fails can one appeal to a mentalistic hypothesis for explanatory help, because, as we shall see later, mentalistic hypotheses, despite these methodological flaws, succeed in providing us with fairly good explanations. (I do not believe that theories that meet the criterion of measurement but not that of consistency are possible. As we shall see below, the criterion of measurement depends on that of consistency.) And now I move on to discuss why these latter hypotheses fail in the two criteria. To answer this question we shall look at several examples.

A theory of the first kind (that meets the criteria of consistency and measurement): an example from physics

Let us look at the explanation that physicists provide for the question, why does a stone fall freely a distance of 4.9 meters in 1 second? The answer, as every high school student knows, is based on Galileo's law, the law of free fall of bodies:

$$D = 1/2GT^2$$

where D is the fall distance, T is time, and G is acceleration of the body as a result of gravity. Now, say the physicists, set 1 second as T and you get D = 1/2G, that is, the fall distance is 4.9 meters. In other words, the stone, like any other body, is drawn to Earth because of Earth's gravitational force, and like any other body the stone in free fall will drop 4.9 meters in 1 second. This then is the explanation; the stone behaves according to a physical law discovered by the renowned Italian scientist Galileo Galilei.

Does this law meet these two criteria (consistency and measurement)? To answer this question we must make the usual distinction (see the above literature) between the language in which the law is formulated and what this language represents. The language of the law is the language of mathematics, which meets the consistency criterion and which describes a certain functional relationship (a square function) between input (time) and output (distance). In addition, the physical terms, and their relations, as presented in the law, are well defined theoretically in the framework of Newtonian physical theory, among other things because these are well linked to the observational level. That is, this law also satisfies the requirements of the measurement criterion. (Although Figure 4.1 shows the diagram of the structure of a theory expressed in behavioral terms, which are relevant to the subject of this book, this diagram may clearly be applied to physical theory by a change of behavioral symbols to those representing physical terms. For example, in the present case on the theoretical level the s variable has to be changed to the T variable, and the r to D.)

That Galileo's law satisfies the measurement criterion calls for an extensive discussion. The explanation for free fall is so standard that we do not even contemplate these questions: on what grounds are we permitted to plug into the theoretical symbol T of this law the observation of time, which we measured by means of a clock? On what grounds are we entitled to place in this law the observation of distance that we measured with a ruler instead of the theoretical symbol D?

In fact, there are several ways to measure time and distance. For example, David has very clear and certain estimates of time and distance: "I have now been waiting for Ruth exactly forty minutes"; "The distance from here to the tower is one hundred and seventy-three and a half meters". Did Galileo mean these measurements? Probably not, but why not? What is so special about measurements with a clock and with a ruler? And why do we not accept David's psychological measurements, estimates, as appropriate? To answer these questions we shall look at the way physicists conduct observations and connect them to the terms in their theory by the procedure of "fundamental measurement" (e.g., Cambell, 1953; Coombs, Dawes, Tversky, 1970; Michell, 1990, 1999).

The theory-observation connection is created by measurement of objects' properties, such as length, weight, and time. Without going into the debate between the various theoretical approaches to such basic terms as property, number, and measurement, I shall describe here "fundamental measurement" as a process in which the relation between a certain quantitative property of an object and the unit of measurement of this property is revealed empirically. For example, if we have before us a straight stick of length X and we find that another stick of length M (which we determined as our unit of measuring, for example, a meter) goes into X from end to end exactly ten times, we have discovered that length X is ten lengths M (that is, X/M=10). Even though at first sight what we have discovered here appears trivial, this procedure has immense significance on which all of physics rests. Why?

The essential point in this measuring is that scientists found an empirical operation (counting how many times M goes into the length of the object measured), which upholds mathematical properties that define the world of numbers on which the mathematical symbols in physical theory are based, for example, D, T, and G in Galileo's law. Let us look at such properties as these two: transitivity and additivity. The transitive relation states, for example, that if 3 < 20 and 1 < 3, then 1 < 20; the additive relation suggests, for example, that 20+3 = 23. Are these relations also maintained in the group of sticks? The answer is affirmative, and we shall demonstrate it by means of the three following sticks:

```
A |--------------------|
B |---------------|
C |-----|
```

As a first step, we shall define the unit of measurement of length by means of the piece (-); as a second step, we shall count how many times this unit goes into A (20 times), B (15 times), and C (5 times); and as a third step we shall see that the lengths of the three sticks indeed uphold the transitive relation, because A is greater than B, B is greater than C, therefore A is greater than C; also the additive relation, because A = B+C (20=15+5).

As the measurement of length upholds all the mathematical properties of numbers, it transpires that *what we say by means of numbers will be said also by means of the lengths of the sticks*. And that, to my mind, is the closest connection it is possible to make between the theoretical term (a symbol that represents numbers) and observation (length of an object). (This fact does not determine that scientific hypotheses expressed in mathematical language are correct. The question of the correctness of a scientific hypothesis depends, of course, on the results of empirical experiments.)

The same may be said about several more quantitative properties of this kind, such as weight and time. (Measurement of weight is based on the use of scales, and measurement of time is based on the use of a periodic phenomenon, for example, Earth's revolution around the sun. Physics uses additional quantitative measurements such as density and temperature. I shall not deal with these, because the theory of measurement which underlies them is not simple and cannot be easily and intuitively presented, like the measurement of length.)

In sum, what is stated in the world of numbers represents precisely the relations in systems of length, weight, and time. Moreover, by means of these measurements of length, weight, and time physics has succeeded in building a great and efficient theoretical system that includes complex concepts such as speed, acceleration, work, and energy, by means of which scientists have been able to understand the world, and change it.

A theory of the second kind (which satisfies the criterion of consistency but not the criterion of measurement): an example from the theory of learning

How do researchers of animal learning explain the phenomenon whereby a hungry rat placed in a Skinner box (or operant box) with two levers presses lever (a) three times more than lever (b)? The answer is based on the use of the matching law or on a law named for Herrnstein (Herrnstein's law: Davison & McCarthy, 1988; Herrnstein, 1961):

$$B_a/(B_a+B_b) = rf_a/(rf_a+rf_b)$$

where B_a (Behavior$_a$) denotes the frequency of presses on lever (a), B_b denotes the frequency of presses on lever (b), rf_a denotes the frequency of reinforcements (food) given for presses on lever (a), and rf_b denotes the frequency of reinforcements (food) given for presses on lever (b).

Now, as becomes clear from the experimental design, the relation between reinforcements is $rf_a = 1/3\ rf_b$, so according to the matching law the relation between presses also will be $B_a = 1/3\ B_b$. And this is the answer to our question above: the rat presses on lever (a) three times more than on lever (b).

Clearly, this law, this theory, meets the criterion of consistency, as the relations between the conceptual symbols are well defined mathematically. Furthermore, in addition to the wide empirical support that this law has gained, highly interesting and successful efforts have been made to ground this law in conceptual frameworks such as economic theory and signal detection theory (see review and discussion in Davison & McCarthy, 1988). Still, as we shall immediately see below, this law does not satisfy the measurement criterion, which also blurs the psychological concepts themselves.

To substantiate this matter, we shall look into the concept of 'reinforcement', which is used in the context of learning, of change of behavior. For example, a hungry rat learns to press on the lever because this response is accompanied by the presentation of food, that is, a change is created in its behavior; it has acquired a new behavior. But clearly, the strength of the reinforcement (to change the behavior) depends on the rat's neurophysiological-mental state. For example, if it is not hungry the reinforcement will have no effect. Furthermore, in a sated animal food becomes a negative stimulus that it is anxious to avoid. The number of hours that the animal is left without food exerts different effects on different species, and on different individuals of the same species. Max, as we saw above, refused to eat other foods that he was not accustomed to, and a similar feature has been found in other cats. Finally, what constitutes reinforcement in one has no value in another, while in a third the same stimulus is perceived as something abhorrent. Contrary to physics, then, an absolute, valid, and reliable definition cannot be suggested for the concept of reinforcement, and it is impossible to point to a unit of measurement of "reinforcivity" with which we can measure the dimensions of "reinforcivity" of various stimuli such as food, water, heat, sex, and dignity.

In light of this I shall raise the following general question: in psychology is the connection between concepts and observations based on the measurement procedure followed in physics as described above? I believe that the answer is negative. To check this, we shall look at what is done in psychology. Psychologists connect concepts and observations by means of an "operational definition", namely the meaning of the concept is obtained by detailing the processes of observation and measurement. For example, learning is defined operationally by counting the problems that were correctly solved through a given training period; aggression is defined by the number of words or motor movements considered aggressive in a certain culture; fear is defined by the distance of flight from the threatening stimulus or the period of remaining crouched; time is defined by subjective evaluation of the physical time span that has passed between two events; distance is defined by estimation of the space between two stimuli located in a space; and the cognitive effort required to activate cognitive processes that are suitable for solving a given problem is defined by the length of time – latency – that it takes to solve this problem.

Do these examples confirm that in psychology too the concept-observation connection is as close as it is in physics? In my opinion they do not, mainly for the following four reasons.

1) *The problem of validity*: to illustrate this problem we shall look, for example, at latency. Does latency indeed measure only the cognitive effort in a manner similar to the concept-observation in physics? The answer is no. Measuring the length of the object with a standard ruler focuses on that alone, and does not measure the color or kind of material out of which the object is made. Latency, by contrast, is affected by a large number of different cognitive factors and processes, such as weariness, excitement, and degree of interest the individual has in the given problem, which have no direct link to cognitive effort. Furthermore, latency does not measure everything connected with this effort, as it is not known which cognitive processes go into action when the individual tries to solve a given problem. For example, which processes (and what speed of their activity) are involved in coding the information of the problem? Does the individual use shortcuts based on past experience? To solve the problem, does the individual need additional information present in her memory, and what is the speed of retrieval of this information? Do these processes work one after the other, or in parallel? In short, latency is not a valid measure of cognitive effort: it does not measure only cognitive effort.
2) *The multi-dimensionality problem*: While the concepts of psychology are multi-dimensional and subject to interpretations from different and varied viewpoints, the concepts of physics (length, weight, time) are uni-dimensional. Furthermore,

while it is very hard to break down the psychological concepts into their uni-dimensional elements (try, for example, to break down the concepts of love and envy), in physics the complex (multi-dimensional) concepts are composed of uni-dimensional concepts. For example, speed is composed of the relation between distance and time; acceleration is based on the relation between distance and time squared; and kinetic energy consists of weight and speed squared.

In psychology, because the concepts are multi-dimensional, in many cases the transitive relationship is breached too: for example, our friend David maintains that Nimroda is more beautiful that Zilpa, because of her green eyes; Zilpa is lovelier that Jocheved because of her raven hair; but he insists that actually Jocheved is more beautiful than Nimroda because of her long fingers.

3) *The problem of private experience*: Like physicists, who ascribe numbers to physical properties, for example, the size of this pole is five meters, psychologists ascribe numbers to psychological properties; for example, on a scale that measures interest from one to ten, David's lecture is eight, the intensity of the note A is perceived as double the intensity of the note B, the number of David's correct answers in a face-recognition test is seventy out of a hundred, and so on. Is the ascription of numbers in psychology endowed with the same qualities of measurement as in physics? In my view it is not. In physics, scientists ascribe numbers to physical properties present in the world, outside the human's cognitive system, but in psychology numbers are ascribed not to behavioral properties present in the world, outside the human's cognitive system, but to stimuli *as the human perceives them, that is, ascription of numbers is a product of processing information that is performed in the human's cognitive system.* In other words, while physical properties to which numbers are ascribed are independent of the observer (thus measurement in physics is objective), the psychological properties of the stimuli to which numbers are ascribed depend on the observer – on her perceptual system.

   For example, let us take another look at the famous illusion of Müller-Lyer: <>—<. The physicist finds that the length of the right-hand line is identical to that of the left-hand line; but the participant in a psychological experiment assigned to the length of the left-hand line a numerical value smaller than the numerical value she assigned to the right-hand line. The reason is this: while the physicist assigns numbers to physical properties outside the perceptual system, that is, finds that the length on the right is equal to the length on the left, the person in the psychological test responds to the right-hand and left-hand stimuli according to the processing of the information that takes place in her perceptual system, and that finds meaningful expression in her mind: the left-hand line is shorter than the right-hand line.

   This argument may be generalized to most responses: our responses (and those of animals) are not in the nature of a purely motor movement, reflexes produced automatically, but are responses bearing meaning, responses that express the processing of meaningful information performed in our consciousness. I raise my arm not as a mere motor movement but in greeting. I walk not because my legs all of a sudden have started to move, but because I want to reach a certain place or because I want to stretch my limbs. In short, our responses, our activities, and our actions are brim-full of intentions, desires, ambitions, sense of awareness, that is, private experiences, and their entire essence is nothing but the private experience.

But a unit of measurement cannot be found for private experience, like that which exists for length. It is impossible, for example, to take the sense of the taste of lemon (not the physical stimulus, the lemon solution) to define the unit of measurement of the taste of lemon (the "meter" of the lemon, M_L) and by means of it to measure the taste of the lemon L, so that we would obtain, for example that $L/M_L = 10$, that is, that the taste of the lemon of L is ten M_L (ten units of lemon taste). It is impossible to take the feeling of love, to define a unit of measurement of love, and say that Jacob loved Rachel by ten and a half units of love more than he loved Leah. And it is impossible to take sharpness of memory of Bialik's poem 'To the Bird' to define a unit of measurement of sharpness of memory, and to say that sharpness of memory of this poem is three and a half times the sharpness of memory of the first year in kindergarten. Suggestions of this kind sound nonsensical. Just as it is impossible to find a unit of measurement for private experiences, so is it impossible to find units of measurement for responses that express these experiences. That is, there is no possibility of measuring these responses, as is customary in physics.

What then do psychologists measure? They do not measure a response bearing meanings but a motor, neurophysiological, response alone: they measure the raising of the arm, and not its meaning; in the Müller-Lyer illusion they measure if the subject says that the left-hand line is shorter than the right-hand, but they do not measure the private experience of perception of the different sizes, the consciousness of this perception, that which is the essence of this response. The reason is exactly the same reason we considered earlier: only the human herself is aware of the variegated meanings of her responses, and she alone mentally experiences what she does and the meaning of the deed.

4) *The problem of the requirement of "unit equivalency"*: While physical theory, the physical law by means of which an explanation is supplied in physics, satisfies the requirement that I call "unit equivalency" (see Rakover, 2002), psychological theory does not. According to the requirement of unit equivalency, the combination of measurement units on the one side of the law's or the theory's equation has to be identical to the combination of the measurement units on the other side of the law or equation. To clarify this requirement, we shall look once more at Galileo's law, the law of free falling of bodies:

$$D = 1/2GT^2$$

where D is distance, T is time, and G is acceleration of the body as a result of the force of gravity.

Now, since D is measured by the unit of the meter, the expression GT^2 likewise has to be measured by the unit of the meter. Sure enough, a simple algebraic calculation shows that it is: meter = [meter/time^2] time^2.

Only through satisfying this requirement, which is based on the procedure of fundamental measurement, is the physical explanation possible that clarifies for us why certain relations are maintained between different physical events, for example, the relation of gravity between Earth and a falling body; and only thus is it possible to understand and calculate how much strength is needed to pull a wagon loaded with sacks of potatoes, and with what force a stone has to be thrown to smash a window. Furthermore, without upholding the demand of unit equivalency it would be impossible to formulate the laws of conservation that are so important for understanding the behavior of different physical systems.

Does psychological theory satisfy the requirement of unit equivalency? It does not. Let us look at the example of the general structure of a theory or a law in psychology:

Behavior = f(Stimuli, Neurophysiological processes, Cognitive processes)

Now let us ask if the units of measurement of behavior (e.g., percentage correct responses, speed of response, force of response) are identical to the combination of units of measurement of the stimuli, the neurophysiology, and the cognitive processes. They are not. The number of correct responses is not identical to the physical units of the brain processes (e.g., differences in electric potential), to the units by which the stimulus is measured (e.g., loudness of the noise), or to the units of measurement of cognitive processes (which in most cases are conceived of as information processing). Even though units of measurement for physical, chemical, and physiological processes exist, it is not possible, as I argued above, to find units of measurement for the meaning of the response or the information processing. Furthermore, the notion of information in psychology is not absolutely defined, and in fact constitutes a kind of hold all into which one can put everything imaginable having to do with knowledge – a notion that defies any attempt at definition (see discussion in Palmer & Kimchi, 1986).

This description does not apply to the matching law alone, but, I believe, to all psychological theories and models formulated in mathematics language or in computer language. Hence, the theory-observation connection in psychology is not based on the theory of measurement employed in physics. So on what is the connection between the concept and the observation based in psychology? I shall not be far off the mark if I say that this connection is based on cultural intuition (e.g., aggression is defined by curses, pushing, and blows, because usually this is how people offended by others respond), empirical knowledge that has accumulated from research in a given field, accepted procedures (this is what all researchers in the field do), and more than a pinch of arbitrariness.

In light of this discussion, it may be suggested that not satisfying the measurement criterion projects lack of clarity onto the world of theoretical concepts. This point is particularly salient when one examines the implications of the requirement of unit equivalency. If psychological theory does not meet this requirement, that is, the connection between its concepts and the observation is very loose, one can arrive at odd theories – that a person's IQ equals the number of pairs of shoes multiplied by the number of pairs of trousers; or that Einstein's IQ (say 150) equals the IQ of three morons (each of whom has an IQ of 50). In brief, an awful situation is liable to arise in which psychological theory will play the mathematics game correctly without its concepts reflecting the processes that take place in the individual.

Theory of the third kind (which does not meet the consistency and measurement criteria): An example from everyday psychology

How may we explain the fact that David traveled in a taxi from Tel Aviv to Jerusalem? The explanation is based on an appeal to David's private behavior: David wanted to get to Jerusalem (to meet Ruth, for example) and believed that a taxi ride would realize his wish. Generally, it is possible to suggest for explanations of this kind a formulation of a kind of "law" called in the literature the "purposive law" or "teleological law" (see discussion in Rosenberg, 1988): if X has a wish, motivation, to achieve a certain goal, and X believes that performing a certain behavior will achieve, realize, this goal, it will be logical for X to perform that behavior.

As the foregoing arguments about not satisfying the measurement criterion apply to the present case too, I shall move directly to a discussion of the question of consistency. This law does not satisfy the requirement of consistency because from the language in which it is formulated no inter-concept connection is obligatory, as it is in the mathematical formulation in the two previous examples. It is hard to see how the language of mathematics can be applied to this law, as an obligatory connection is not sustained (nor a probabilistic one either) between the concepts of desire, belief, and behavior. The behavior does not stem with logical, mathematical, necessity from desire and from belief; and there is no physical necessity here, as there is bound to be by the law of free fall, the law of Galileo. The relationship between these concepts is practical, grounded in daily life. It would come as no great surprise if it turned out that although David wished to reach Jerusalem, and even believed that traveling in a cab would make his wish come true, he did not put his intention into practice. Ultimately, this behavior (a journey to Jerusalem) is in the nature of David's free will. (Later, especially in chapter 7, I shall discuss whether it is possible to regard teleological law as a law or a theory, as these terms prevail in the natural sciences.)

As may be seen from this discussion, Galileo's law sets before us an example of a perfect mechanistic explanation: it satisfies the two criteria of consistency and measurement. (These two criteria are not the only ones for characterizing mechanistic theory, but they are enough to create the distinctions that are the matter of this chapter.) Matching law also is an example of a mechanistic explanation, and it seems to meet the criterion of consistency, but it fails, like the other theories in psychology (including cognitive models based on analogy with the computer), in the measurement criterion. Despite great efforts, the development of behaviorist concepts free of reference to the internal, mental word, I believe, has not succeeded. Teleological law offers us an example of a mentalistic explanation, an explanation that resorts to the individual's internal world, but that fails on both counts: on the criteria of consistency and on that of measurement.

From this aspect, if we accept these criteria as reflecting what is considered science in our eyes, our order of preference for giving an explanation for a given phenomenon must be the following: theories of the first kind that meet the criteria of consistency and measurement; next, theories of the second kind that meet the consistency criterion but not the measurement criterion; finally, theories of the third kind that do not meet the consistency criterion and the measurement criterion.

Given this analysis of kinds of theories in science, the following question arises: what harm is there in using theories of the third kind? Here is the answer: not meeting the consistency criterion is liable to impair the ability to produce predictions, and not meeting the measurement criterion is liable to impair the ability to connect observation to prediction, that is, the capacity for the empirical test. These impairments are liable to generate serious doubts as to the conclusion that arises from the results of the empirical test, and on the other hand they are liable to allow very easily the suggestion of ad hoc hypotheses intended to save our theory.

This analysis gives rise not to a proposal that the mentalistic hypothesis (e.g., the anthropomorphist hypothesis) should be seen as a sore evil, an idea disqualified from the outset, but only to the suggestion that it should be seen on the one hand as a hypothesis that can be put to the empirical test, but on the other hand as a hypothesis that requires extreme caution in scientific treatment, on account of the above weaknesses. This stance does not accord with that of other researchers. For example, in Kennedy's (1992) opinion, Thorpe's hypothesis that birds' building a nest is a purposive activity did not win empirical support, for the simple reason that this hypothesis is anthropomorphist: "It is again noteworthy that none of the authors suggested why Thorpe's idea could not be confirmed, although the reason is plain: it was an anthropomorphic idea" (p. 37).

4.2 Why should one use a mentalistic explanation?

The answer is plain: because there is a large group of behaviors in animals that can hardly be explained by an appeal to mechanistic theories (i.e., theories of the first and second kinds described above). Although here I shall concentrate on the behavior of Max the cat, it is worth noting that the literature offers many examples of intelligent behavior in animals, which attests to thought and awareness, behavior that is very hard to interpret just by means of mechanistic explanations. For example, Levy and Levy (2002) write:

> Studies on animals have come a long way since humans learned to communicate with them, and gone is the view that animals' behavior is dictated solely by instincts, and that superior animals (mammals, birds) are able, at most, to exhibit a limited capacity for learning.... Later we shall try to show that animals have consciousness and also a degree of self-consciousness, and that they are able to solve problems in elaborate ways and by means of amazing insights across a considerable variety of areas. All these undoubtedly attest to intelligence, and not only to intelligence, and the chief thing is that they reinforce our basic assumption, which passes like a golden thread through the entire book, that animals should not be seen as 'things' or objects, but as 'persons'. (p. 16)

We shall look at several examples that invite mentalistic interpretations (despite the criticism: see Levy and Levy): the behavior of rodents and birds that store food in spring for consumption in winter, a behavior that testifies to purpose and planning in advance; behavior of intentional deception (birds limping) and feigning death; adaptation to a human environment by crows in Japan, which learned to utilize the movement of cars to crack nuts that the birds placed on a pedestrian crossing beforehand; the ability in higher primates and in cormorants to count; language learning, and using it and various signals to create communication in parrots, chimpanzees, dolphins, and whales; animals' ability to express their emotions artistically and to enjoy artistic creativity (drawing in monkeys, singing in birds); mourning behavior in elephants following the death of one of their kind; empathic behavior toward the suffering of one of their kind in dolphins, elephants, dogs, and primates; and finally, cooperation between fishermen and dolphins, which chase fish inshore where the fishermen have previously spread their nets (and see Allen & Bekoff, 1997; Griffin, 1981, 2001).

Now we shall examine two groups of behaviors of Max the cat: mechanistic behavior and mentalistic behavior.

Mechanistic behavior includes the following group of behaviors (based mainly on Morris (1997)), which were observed also in Max: fur-licking, fear-aggression response (arching the back, hissing sounds), landing on all four paws, sharpening claws, ear twitching, tail wagging in a situation of decision, excretion (urinating, defecating), mating, play and hunting, rejection of food, social behavior (principally with humans, for example, prolonged gazing at a human), beating (milking) movements of the fore-paws, and erecting rear and tail in response to movements of combing and stroking the back. All these are characteristic responses in cats, and their explanation is through an appeal to hereditary-evolutionary neurophysiological factors.

According to the renowned article by Tinbergen (1963) there are four types of explanation, which together offer a whole understanding of animals' behavior. One is the nearby causal explanation, which refers to factors close in time and space to the behavior under study, such as perception, cognitive representation, decision-making, and neurophysiological processes. Next is the developmental cause, referring to the individual's genetic background and past experience. Third is the functional reason explanation, which refers to the contribution of behavior to the individual's adaptive-survival ability. Finally there is the evolutionary cause, dealing with evolutionary factors that have shaped the response under consideration (and see discussion in Hogan, 1994; Shettleworth, 1998). These explanations, clearly, are nothing but different kinds of mechanistic explanations.

Mentalistic behavior covers most everyday behavioral episodes (in Max's living space in the Rakover family's apartment) described above: **ambush for the night moth, ambush for Aviva, head on cushion, wailing-vocalization, non-change of direction of walking, visit to the vet, armchair-catch game, armchair-petting game, tail game, jumping onto the knees, getting off the knees, armchair-knees scratching, tickling the neck, combing fur, continuing with the petting, continuing with getting off the knees, continuing jumping onto the sofa, being seated on the knees, non-change of lying, preference for Aviva, sleeping: running-rolling over onto the back, sitting before jumping,** and **entering the bedroom.**

All these are episodes-responses that are different from the responses described in the first group. Their explanation does not accord only with the appeal to hereditary-evolutionary neurophysiological factors, and calls for mentalistic explanations, as I showed by analyzing a large number of these episodes earlier.

To drive home this explanatory distinction, I shall compare two behavioral criteria, by means of which a given behavior can be classified into one of these two groups: a criterion for mechanistic behavior and a criterion for mentalistic behavior. The former is based conceptually on (a) understanding behaviors of the following kind: reflex, instinct, fixed action patterns, innate releasing mechanism, innate behavior, and motivational behavior (see reviews and discussions in Frank, 1984; Alcock, 1998; Barnett, 1998; Haraway, 1998; Hogan, 1998; McFarland, 1999); (b) analogy to the action of a machine: a machine responds in the same behavioral pattern when it is in the same stimulus pattern, for example, an air-conditioner, a car, a guided missile, a robot, or a computer. This what Tavolga (1969) writes:

> Traditionally, an instinct is thought of as an automatic response, an inherited or innate behavior, an unlearned behavior that is built-in as part of the structure of the organism. In performing an instinctive act, the animal is considered to behave like a machine that can do only the things is built to do. (p. 91)

"The criterion for mechanistic behavior":
If an ***animal of a given species*** *is in a state of a* ***certain physiological drive*** *and in an environment of* ***relevant stimuli for this drive, a fixed and characteristic pattern of responses*** *will awaken in it.*

The words in boldface have the status of a variable that obtains different values. The term **animal** has a status akin to a variable, because it can include names referring to different animals of different varieties, such as rats, cats, and different kinds of cats. The term **certain physiological drive** serves as a category that can contain different drives, such as hunger, thirst, pain, fear, and sex. The category **relevant stimuli for this drive** includes stimuli such as bread and water, which match hunger and thirst. The term **fixed and characteristic pattern of responses** refers to fixed behavior patterns that appear at different times, in the same individual, in the same species (in different varieties of animals), for example, behavior of hunger, thirst, aggression, and sex.

Now we shall examine two examples that satisfy this criterion.

Example 1. When the hunger drive awakes in a wild cat it begins to wander and look for prey. And when it sees a mouse it performs a pattern of hunting responses, which concludes with the killing and eating of the mouse.

Example 2. Hungry seagull chicks tap their beaks on a red patch found on the lower extremity of the parent's beak, in order to get food which is located inside the beak (according to Frank, 1984).

These two examples involve the same physiological drive: hunger. When the cat is hungry it becomes restless, and goes in search of prey. When the appropriate stimulus (the prey) is perceived by the cat, the characteristic response pattern associated with hunting is activated. And in the presence of a beak with a noticeable red patch, the hungry chicks hammer on the parent's beak to release the food from it. The chicks' interest in the beak will fade when the red patch does not appear on it.

Having clarified what we mean by this criterion, let us test whether Max's behavior satisfies it. As may be seen, all the behaviors that appear under the heading *mechanistic behavior* do so. As an example we shall take the beating (milking) response.

When Max gets onto Aviva's stomach, or mine, he begins to beat with his forepaws in rhythmic movements. This beating is characteristic of kittens suckling at their mother cat's teats, and it promotes the flow of milk into the nursling's mouth (like milking). In the present case the human abdomen arouses in the cat this suckling response pattern (see Morris, 1986, 1997).

The second criterion, that of mentalistic behavior, is based on the conditions for behavior of free will, which I described in the last chapter, conditions that do not accord with mechanistic behavior.

"The criterion of mentalistic behavior (the Free will (FW) condition)":

The same individual, whose private behavior has changed, in the same state of stimuli, at different times, responds ***with different responses*** to realize the ***same goal*** or with the ***same response*** to realize ***different goals***.

This criterion evidently runs counter to the foregoing condition for mechanistic behavior. When a given behavior satisfies the requirements of the "criterion for mechanistic behavior" it emerges that this behavior does not satisfy the requirements of the FW condition, and the reverse. For example, if the same behavioral pattern recurs in the same individual (and in others of the same species), in the same state of stimulus, and at different times, it will be very hard to accept that this behavior attests to free will. By comparison, a behavior that conforms to the mentalistic criterion is explained on the basis of the hypothesis regarding changes in private behavior, in mental states. For example, in chapter 2 above I analyzed the episode of the **ambush for the night moth** and I showed that this episode could not be explained only based on a mechanistic explanation. In chapter 3 I analyzed a large number of behavioral episodes (which figure in the list of mentalistic behaviors) and I showed that they satisfy the behavioral criterion of free will; I deduced by analogy that these behaviors are performed through consciousness. These behaviors, then, are hard to explain as mechanistic, and they call for a mentalistic explanation.

Here the following question arises: are these two criteria really contrary to each other? Is it not possible ultimately to reduce the mentalistic criterion to the mechanistic criterion? Is it not possible to construct a robot cat, "Robocat", of astonishing sophistication, that will imitate the behavior of a cat in full, for example, that of Max, such that Robocat will perform mechanistic and mentalistic behaviors alike? Should the construction of such a Robocat prove possible, it will transpire that the explanation for the behavior will of course be purely mechanistic, and we will have no need for a mentalistic explanation. I believe that building such a Robocat is impossible: this is a complicated subject that I shall discuss in the following chapters (see especially chapter 8). In fact, as will become clear, the philosophy of mind considers such a possibility in the context of the debate about the mind and human consciousness: if it is possible to build a human-like robot, which will do everything a human does and is able to do (perhaps even write a book entitled *To Understand a Cat: Methodology and Philosophy*) then the whole world will be given a mechanistic explanation, including phenomena of mind and consciousness.

Meanwhile, until I start on that elaborate subject, I shall expand the discussion of the criterion of mechanistic behavior, and I shall consider four interesting ramifications, which will help us classify a given behavior for a mechanistic/mentalistic explanation.

(1) In the presence of the appropriate stimulus mechanistic behavior will appear even when the individual is exposed to this stimulus for the first time.

I shall discuss two behavioral episodes that appeared complete in Max from the very first time, even though the cat had never learned to perform them. (As I explained earlier, Max first came to our house at the age of three months and we never saw him learn to accomplish this behavior.)

A. **The falling response**: When Max was about a year old I decided to test if it was true that cats always land on four legs. As I was not sure that this Max would indeed land on four legs, and not on his head or his back, I dropped him onto the bed. I lifted him up upside-down and let go. Max twisted his body while falling and landed on his four legs. I repeated this action several times, and I managed to discern that first he turned his head over (ears above, chin below) and immediately after that he turned, in accordance with his head, the front part of his body and after it the rear part of his body, and just before landing Max arched his back and stretched out his legs. Later I saw a series of photographs of a cat during a fall, and found that Max's falling response was very similar to the falling responses of the cat in the photos (see Taylor, 1986).

Most probably then, the falling response is instinctive and appeared complete from the very first time in our friend Max.

B. **Courtship and mating**: When Max was about four I decided that the time had come for this cat to lose his virginity. One day a friend of my son Omer announced that his female Siamese cat was in heat for the first time. The friend brought the cat, which at once began running around the house with her tail upright, while Max chased her wherever she went. It's on, I said, with big smile. Their affair went on for five days, divided into three days of courtship and two of couplings. Max sang to the Siamese cat a kind of strange love refrain, which he emitted over and over: *mihharroow*, beginning on a low note and rising high. He would circle her at a fixed radius, not taking his eyes off her and not ceasing to croon his *mélodie d'amour* (which began to get on my nerves, to tell the truth). Something surprising happened when the Siamese cat looked at him: Max froze on the spot, even in the middle of a movement. I was amazed to see Max standing on three legs, now with the front leg lifted up, and now the rear one, and his head tilting away from her. When he froze, the singing stopped too (thank God). Sometimes the Siamese cat would leap onto the table in the porch and Max would immediately place himself on one of the chairs under the table, constantly singing to her this weird love song of his. Occasionally, when she caught him approaching her, she would whip out her forepaws and hiss – *kkhsss*. One evening while we were watching TV, a bloodcurdling scream was heard from the bathroom, and I rushed to see what had happened. I found Max sitting in his sphinx posture, his eyes fixed on the Siamese cat, who was lying supine, rolling right and left, and rubbing her back against the floor non-stop. "I think the couple have lost their virginity", I said to Aviva, with a broad grin. Sure enough, I saw them mating that marvelous evening in front of the television. Suddenly the Siamese cat raised her rear and stamped with her rear paws. Max mounted her, and held her by the nape with his teeth, and the Siamese cat moved her tail to the left. A little later Max quickly got off her and the Siamese cat let out a long wail and struck out at Max's chest with her forepaws. I don't think she managed to hit him, because this Max remained sitting in his sphinx posture, looking with a kind of concentrated suavity at the female writhing on her back. About two days later, while I was shaving in the bathroom, the screeching of the Siamese cat was under my feet and she cut open a scratch down the calf of my left leg. In the morning I am a man in a foul mood, short-tempered, and this mating game was driving me nuts. What this cat has done he has done, and done enough, I said in a fury, and I requested Omer's pal to be so kind as to remove his Siamese cat – the honeymoon was over. I found a very similar description of courtship and mating in books about cats (see Taylor, 1986; Morris, 1986, 1997). So presumably, the courtship-mating response pattern is instinctive and it appeared complete from the first time in our debonair Max, as well as in the Siamese cat, the erstwhile virgin.

(2) Mechanistic behavior will appear whenever the appropriate stimulus is presented, and when the degree of influence of environmental factors on it is slight.

The cat will land on four legs in any situation, and will court-mate with a female in heat in various conditions of the environment and in different situations. The importance of changes in the environmental stimuli is minor. What determines matters is the individual's being in the critical state of stimulus: the feeling of the body falling upside-down; the smell and behavior of the female cat in heat. These are the stimuli that arouse in the cat the appropriate innate pattern of responses. And as long as the cat is not faced with an extreme situation that prevents performance of these response patterns, they will appear in different cats and in different environmental circumstances. For example, Max courted the Siamese cat all over our apartment and mated with her in the bathroom, the kitchen, the guest-room, in front of the television, in our presence, and as noted, even under my feet. As another example, let us look at the following behavioral episode.

Max spends a good part of his free time licking himself, and if I interrupt this licking action by holding his head with my hand for several seconds, Max goes right back to it the moment I let go. That is, non-extreme environmental changes do not alter or stop the instinctive pattern of action.

Here it should be stressed that not all the cat's behaviors are performed so mechanically. Some behavior patterns require learning and skill acquisition until they are manifested in their full glory and grandeur. For example the pattern of the hunting response develops through a complex process based on innate components (such as biting through the prey's nape) and on long learning in which the mother cat imparts to her offspring the skills of the feline hunt (see Morris,1986, 1997).

(3) "A scale of mechanistic behavior".

It may be suggested that the degree of our certainty that a given behavior requires a mechanistic explanation grows in accordance with the following scale of properties:

A. *A fixed response pattern that appears in the same individual at different times.* If the behavior changes in different ways over time, it follows that this behavior is not mechanistic. By analogy, we do not expect a car to behave differently every hour in the same situations. If the car does not behave in the same way, doesn't start in the morning, it means that something has gone wrong. If the computer screen shows peculiar signs, it is likely that the poor thing has caught a virus. This condition is necessary but not sufficient, because the fixed behavior pattern may also arise from learning. Habits, as everyone knows, are acquired behaviors that are very hard to change. That is why it is so difficult to decide if a certain fixed behavior is innate or learned, even if it recurs in the same individual, in the same cat.

B. *A fixed response pattern that occurs at different times in the same individual, and in the species to which the individual belongs.* If we find that a fixed behavioral pattern appears in other individuals that belong to the same species, the likelihood that this behavior is innate rises. It is very hard to assume that all cats learned to court and mate in the same way as that observed in our friend Max. And the reverse: it is hard to believe that any alley cat will scratch the armchair of the Rakover household so that the householder will lift him onto his knees and pet him. This is a behavior typical of Max, a behavior that he developed in his living space: the Rakover family and their apartment. By contrast, responses such as the **falling response, courtship and mating, licking the fur** are characteristic responses of all cats, so the probability becomes even higher that these responses are mechanistic.

C. *A fixed response pattern that appears at different times, in the same individual, in the species to which the individual belongs, and in the family to which this species belongs.* If it is found that a fixed behavior pattern appears in an individual, in the species of that individual, and in the overall family to which this species belongs, the probability that this behavior is innate increases very much. It is hard to assume that all cats, tigers, and lions learned to court and mate in the same way as was observed in our Max. It is highly reasonable to assume that an evolutionary process is responsible for this behavior in the family of cats across all its many branches.

D. *A fixed response pattern that appears at different times, in the same individual, in the species to which the individual belongs, in the family to which this species belongs, and in other different kinds.* For example, yawning and dreams: Max usually yawns with his mouth open to the full, exactly like cats, lions, tigers, dogs, and ourselves, humans. Max also dreams. When he was sleeping I discovered the appearance of facial grimaces, mouth trembling, and whiskers twitching, and trembling and flexing of his legs. I have seen such a pattern in dogs and of course in people. People who wake at this stage of sleeping report dreaming. (This well known sleep is called Rapid Eye Movements (REM) sleep.) What does Max dream about? I have no idea. It is logical to assume that this kind of behavior is innate and universal, and hard to explain as learned. It is entirely reasonable to assume that evolution is responsible for this quality in different species, in families of species, and in different kinds.

(4) Criterion for mechanistic behavior and neurophysiological behavior.

The two above criteria (for mechanistic and mentalistic behavior) were based on observations of public behavior and on theoretical concepts (private behavior, mental states) which were interpreted from the observations as the best explanatory hypotheses. So far, by the nature of observation of Max, I have not focused on neurophysiological observations. I have no doubt that these observations are likely to help greatly in deciding if a given behavior requires a mechanistic explanation or not. As an example, I shall discuss the response pattern of aggression in the cat, which is produced not by an ordinary external stimulus, for example, the threatening appearance of a dog, but by passing an electric stimulus through micro-electrodes implanted in the cat's brain (see discussion in Adams, 1979; Flynn, 1972). (This behavior was classified above as mechanistic: responses of fear-aggression (arching the back, hissing sounds).) If the behavior is mechanistic, there is some innate neurophysiological mechanism (in the brain and nervous system) which in the appropriate stimulus conditions produces the behavior independent of the individual's will.

In a series of experiments on a cat, Flynn (1972) showed that in the presence of a rat it is possible to produce in the cat an aggression response by sending the adequate current (in the ranges of .10-.90 ma) through the hypothalamus. As the current in the brain is greater, the intensity of the aggression response is greater. The aggression response was discovered as a reflexive pattern consisting of a chain of reflexes: the cat struck the rat, placed its paw on it, leapt, and bit it on the back of the neck. Without the passage of an electric current, the cat did not attack the rat. Without the presence of the rat the passage of the current in the brain caused the cat to move around the cage sniffing, and show an emotional response of anger including facial expressions and vocalization. Through these and other experiments it was possible to isolate the neurophysiological system in the brain that is involved in the aggression response in the cat. Furthermore, in a review article covering a large collection of studies on the rat and the cat, Adams (1979) posited brain mechanisms for responses of attack, defense, and submission.

As may be seen from this short account of the cat's behavior, the aggression response investigated by means of brain stimulation satisfies the criterion for mechanistic behavior: the aggression response pattern appeared whenever the cat's brain was stimulated by a suitable electric current in the presence of the rat. In fact, the debate going on in the literature does not center on whether it is possible to explain this behavior by means of mentalistic explanations, but principally on the degree of accuracy of the mechanistic explanation – accuracy in the description of the neurophysiological brain mechanism responsible for this behavior pattern.

CHAPTER 5

Three-stage interpretation

The purpose of this chapter is to develop an explanatory procedure for complex behavior constructed out of mechanistic and mentalistic behavioral components. This procedure, called the "three-stage interpretation", breaks down the behavior into its components and focuses on a main one, which constitutes the pivot of the interpretation. In the first stage of the interpretation, the purposive mentalistic explanation is applied to this behavioral component; in the second stage an explanation for this component is proposed according to its typical standard function (e.g., a cat's extension of its claws according to the survival function); in the third stage it is shown that this behavioral element has changed its function and has acquired a new purpose that suits the purposive explanation suggested in the first stage. This interpretation is illustrated by the analysis of several behavioral episodes of Max the cat, and is juxtaposed against the possibility of proposing a mechanistic explanation for these behaviors. In addition, certain differences are discussed between this explanation and other methods of explaining complex behavior by breaking them down into behavioral components.

The sure scientific way to understand behavior is to decompose it into its components; and after I have decomposed, divided, split, and separated, I have reached an infinite void between two theoretical concepts.

As may be seen from the previous chapters, Max's behavior is complex and built of a combination of mechanistic and mentalistic behavioral components. Each and every action of his, in fact, is based on a collection of dynamic processes: physical, chemical, neurophysiological, motor, cognitive (analogous to the actions of the computer), and consciously mental (will, belief, purpose, intention): complex processes that in the end are responsible for the behavior we observe in everyday life.

I believe that this behavioral complexity finds expression also in research with animals. Historically, a line can be discerned in this research, developing from an attempt to explain the behavior of animals on a hereditary evolutionary basis (e.g., Lorenz, 1950, 1965, and Tinbergen, 1951/1969), through emphasis on complex interaction between genetic and environmental factors such as learning, to the understanding that a full explanation of animal behavior also requires reference to mental processes (see discussion in Allen & Bekoff, 1997; Griffin, 1976; Jamieson & Bekoff, 1992). For example, Johnson (1972), who summarizes a rich literature of aggression in animals and humans, suggests that an electric stimulus in the brain arouses an innate neurophysiological system that constitutes only a necessary part in the pattern of response of attack, while appropriate environmental stimuli (such as a rat to a cat) and the animal's past experience are of great importance in performing this response. He writes: "Attack is one kind of species-typical consummatory behavior, and it is characterized neither by total rigidity nor by complete plasticity" (p. 75). According to Tinbergen et al. (1965) the development of behavior is the outcome of a complicated interaction between innate and learnt components.

As an example of behavior that involves innate, learnt, and mental factors, we shall examine once more the episode of the **ambush for the night moth**. I maintain that the following behavioral components that enter into the moth episode may be interpreted as innate components in hunting behavior: staring at stimuli making rapid movement (the moths that circle round the porch light); decline in interest when the interesting stimulus (making rapid movement) stops or disappears; the search for the moving stimulus that has suddenly disappeared; pursuit and jabbing at the rapidly moving stimulus with the forepaws. As against this, the following behaviors are interpretable on the basis of learnt, cognitive, and mentalistic processes: using the information that night moths generally spin around the lit porch light to make a practical inference, which results in setting the ambush for the night moth on the table, closer to the porch light with the purpose of continuing with the amusement.

Hence, the explanation for this episode and its like is based on behavioral components that require use of two kinds of explanation: mechanistic and mentalistic. This explanation gives rise to the central question of this chapter, and the following ones: What is the nature of an explanation of this kind?

The answer will be given in two parts: in the first part, in the present chapter, I concentrate on the explanatory design, which I call the "three-stage interpretation", which focuses on a principal mentalistic behavioral component; in the second part, in the following chapters, I shall concentrate on a "multi-explanation theory" which treats a large number of mentalistic and mechanistic behavioral components.

5.1 Three-stage interpretation and the principle of new application

The three-stage interpretation is based on a match between the purposive mentalistic explanation (will/belief) and the Principle of New Application, which sets two conditions that attest to mental processes:

a. Use of an existing response to achieve a new purpose, different from the earlier function (adaptive-survival) of this response;
b. Application of different behaviors (including new responses) to achieve the same purpose.

The match is performed by identification of the new purpose with the purpose of the mentalistic explanation or by realization of the purpose of the mentalistic explanation by means of different responses.

Before I proceed to discuss in detail the three-stage interpretation, it is important to consider the following question: what is the point in suggesting basing the three-stage interpretation on the purposive explanation? Mentalistic explanations, which appeal to internal factors to explain a given behavior, are not restricted solely to the purposive explanation (will/belief) and in fact they appear in various forms and kinds, for example, explanations based on following rules of behavior, cognitive ability (thinking and inferences), feelings and emotions. Furthermore, in most cases an explanation of behavior is based on several kinds of mentalistic explanations. For example, the episode of the **ambush for the night moth** is explained both by Max's wish to continue the chase after the moth and by an appeal to his cognitive ability to understand that the moth may resume its flight around the porch light.

The answer to this question lies, on the one hand, in the great importance that researchers attach in their discussions to the purposive explanation (will/belief) in cognitive and folk psychology, and in explanations of action. In fact, the explanation by means of will/belief is the main one, if not the only one, that features in these discussions (see discussion on the subject in chapters 7, 8, and 9); and on the other hand, as we shall see later it transpires that the purposive explanation is appropriate for a considerable part of Max's mentalistic behaviors (see chapter 6).

As an example of the great importance of the explanation by means of will/belief we shall study the conceptual similarity between the purposive explanation (will/belief) and the explanation by means of rule-following (e.g., Rakover, 1990, 1997).

Rule-following: When Uri stops his car at a red light we explain the central behavioral component in the complex behavior of driving – pressing on the brakes pedal – as obedience to the highway code: as a good and disciplined driver, Uri has followed the rules of traffic and has performed a stop at a red light. In the same way, it is possible to explain part of Max's behavior by means of several rules of behavior that pertain in the Rakover household. For example, Aviva does not want Max to lie on her bed so she has taught him the rule of "No Entry into the Bedroom". (The cat is a very clever animal and it is enough to scold him once or twice and to close the bedroom door for him to get the message very well. However, see infringing prohibitions in chapter 3.) Only thus is it possible to understand the fact that Max customarily waits for Aviva at her bedroom doorway when the door is open. Furthermore, if Aviva is lying on her bed reading, Max usually draws her attention to the fact that he, His Catship, is present here at the threshold, through a plaintive miaow. (The musical variations of the sound that Max is capable of producing from his throat, I believe, are an interesting subject in themselves.) And sometimes, when Max's patience is at an end, he creeps stealthily into the room. But then one irate word – "Maax" – is enough for the cat to turn tail with a short snort of exasperation.

Purposive explanation: When David wants to read a particular entry in the encyclopedia he stands a chair to take the appropriate volume off the top shelf. We explain the manifestation of this behavioral component – climbing onto the chair – by appealing to the fact that in David's view this behavior is likely to realize his purpose. A large part of Max's behavior can be explained in the same way. For example, in the episode of **scratching the armchair-knees** Max wants me to pet him, so to catch my attention he approaches the armchair in which I'm sitting and scratches the seat with his claws. I bend towards him, place him on my knees, and stroke him. (Sometimes, after a while, Max also begins beating my stomach.) Occasionally he stands before me and rolls over onto his back, presenting me with his white belly, and fixes my gaze with his blue eyes. And again, I bend over him, sometimes to tickle him on the belly and under his jaw, and sometimes I set him on my knees (see **rolling onto the back: continue petting**).

These two kinds of explanation are close to each other because in many cases, on the one hand rule following is done for the sake of a clear goal, and on the other hand the will/belief is realized through employing rules of behavior. For example, David wants to meet Ruth in Tel Aviv and believes that driving according to the highway code from Haifa to Tel Aviv will realize his wish; and Max acts according to the rule of "No Entry into the Bedroom" and waits at Aviva's bedroom doorway so as to be close to Aviva, and at the same time to avoid her scolding. (Note nevertheless that there is no full overlap between these two kinds of explanation; for example, learning a new behavior for the purpose of achieving a goal by its very essence is not a kind of following rules of behavior. See discussion also in chapter 6.)

Now having clarified this matter, I shall move on to discuss the episode of **scratching the armchair-knees** to illustrate the three-stage interpretation. The specific purposive explanation of this episode is based on the general scheme of the purposive explanation:

If *X* wishes to achieve *G* and believes that behavior *B* will achieve *G*, it is reasonable that *X* will do *B*.

So all we have to do now is replace the capital letters above by the case connected with Max the cat:

Max wants to achieve *petting from me* and believes that *scratching* the armchair will achieve *petting* (*will cause me to pet him*), therefore *Max scratches* the armchair.

But here, before I discuss the explanatory aspects of this behavior, I must consider the complex connection between intentionality and consciousness.

This explanation assumes that one may ascribe to Max conscious mental states similar to what we ascribe to humans in terms of intention, will, belief, and purpose, even though it is hard to assume that Max has a perfected language as humans have. Is this ascription justified? I believe so. Chapter 3 suggested a basis for its justification: different actions in Max satisfied the criterion of free will connected with consciousness, that is, these responses were made consciously by Max. Consciousness does not feature in all cognitive processes involved in these actions (such as retrieving information from long-term memory), but chiefly in the end results of these processes (e.g., consciousness of the content of thoughts, of images in the head, after these have been retrieved from long-term memory which is hidden from our awareness). These matters, as may be seen, are quite complicated, and therefore require several brief clarifications (and see further discussion of consciousness in chapter 8).

While in daily life the term "intentionality" is directly connected with the term "purpose", in philosophy the term "intention" has a broader meaning, which refers to a quality of mental states. These states have semantic significance: they represent content, they represent other states or events. That is, intentionality is a mental quality that allows an appeal to or about mental representations. Thoughts are about something, they refer to a certain content represented in the brain whether this is real or imaginary (see discussion in Allen, 1995; Allen & Bekoff, 1997; Crane, 1991; Rakover, 1990; Searle, 1992; Tye, 1996).

One of the interesting questions linked to intentionality is the following: is intention conscious? At first sight the answer would be affirmative: my thought circles around some subject and I am conscious not only of this subject but also of the fact of my being conscious of this consciousness. But it isn't quite so simple. Let us look at the following example: David wants to meet Ruth, so he travels by train from Haifa to Tel Aviv. The train ride takes about an hour, and during this time David reads a gripping thriller and does not think about the purpose of his journey even for an instant; however, the moment the train stops his wish to meet Ruth returns, and settles in his mind. Hence, David's intention was in his mind before the journey, disappears from it on the journey, and returns to it at the end of the journey. Making the very reasonable assumption that David's intention is not erased from his brain during the journey, because it comes back to his mind immediately the train stops, we reach the conclusion that unconscious intentions exist. This example is not the only one of unconscious knowledge. In research, psychology has discovered different kinds of unconscious wishes, beliefs, and knowledge that have a great effect on behavior. Examples are unconscious urges, uncovered by psychoanalysis, which are liable to cause abnormal behavior; much knowledge that guides us to do things without our being conscious of every responsive component in the act (e.g., playing the piano or violin; riding a horse, or driving long distances), an innate information structure by whose means we acquire language from the moment of birth; important scraps of knowledge that direct our behavior without our being conscious of them (who, for example, thinks during a meal that objects are hard, and therefore the soup does not drip from the spoon?). Also, much research evidence exists of implicit learning and memory in brain-damaged and normal people (see, e.g., discussion in Rakover, 1993; Schacter, 1989).

Philosophers are divided over the question of the link between consciousness and intentionality (see discussion in Allen & Bekoff, 1995; Horgan & Tienson, 2002; Tye, 1996). For example, Horgan & Tienson and Tye hold that the two should not be separated, and that consciousness has intentionality content; but Millikan (1984) suggests that a distinction should be made between consciousness and intentionality, which constitutes, in her opinion, a biological property that has developed by evolutionary processes; McGinn (1982) suggests that the feeling of pain has no representative content, as perception has; and Rosenthal (1991) offers evidence supporting the argument that consciousness and "sensory quality" are independent.

A further important question here concerns relations between intention and language in animals. Several philosophers maintain that language is a condition without which intentionality is not possible, and therefore it is impossible to describe and explain behavior of animals by ascribing to them mental states and processes such as intention, will/belief (e.g., Davidson, 1975; Dennett, 1969; Stich, 1983; and see discussions on this in Allen, 1995, 1992; Allen & Bekoff, 1997). For example, Allen (1992) argues, against Dennett and Stich, that despite the difficulty it is possible through the English language to approach a reasonable description of the belief of an animal (a dog) and to suggest a suitable intentional explanation for its behavior. I agree with Allen, and argue further that it is possible to formulate a hypothesis, which is empirically testable, that intention, will/belief, and consciousness find expression not only in language but also in other behaviors different from linguistic behavior. In general, then, I suggest the hypothesis that animals (such as Max the cat) are creatures some of whose actions are performed out of consciousness and will/belief, through their very possession of consciousness in varying degree.

However, even though I regard living creatures (humans and animals) as having consciousness, I accept that there are mental processes and states of intention and will/belief that may be found outside consciousness. Yet I suggest as a general hypothesis that most cases connected with actions in everyday life occur out of conscious intention in the sense that emerges from the above example, and the following one. David goes into the kitchen with the purpose of making himself a cup of coffee. But on the way to the kitchen he becomes preoccupied with the mind/body question, which makes him forget his intention, so he stops in the middle of the kitchen without understanding what he is doing there. He has already left the kitchen when he suddenly remembers his intention, so he goes back and makes his coffee. Hence it arises that if an intention to act in a certain way is not present in the consciousness (e.g., it has become forgotten, is in long-term storage), the intention will not be realized – the action will not be performed. We reached the same conclusion with the earlier story of the train too: it is reasonable to assume that had David forgotten his intention he would not have gone to meet Ruth. David then would be sitting there on a station bench in Tel Aviv in great despair, and asking himself, what the devil am I doing here? This hypothesis of a connection between everyday intentionality and consciousness goes together with what Searle (1992) had to say: "Conscious states always have content. One can never just be conscious, rather when one is conscious, there must be an answer to the question, 'What is one conscious of?'" (p. 84); and with the assertion of Allen & Bekoff (1997): "When beliefs are attributed to human beings, they are taken to be conscious mental states with semantic content" (p. 66).

By this, I do not state that intention can be realized or influence behavior only if it is present in consciousness. Experimental results in implicit learning and memory show that implicit knowledge, knowledge that the individual is not conscious of, exerts an effect on her behavior, although – such is my impression from this collection of experiments – the influence of the unconscious on behavior is different from (and in many cases its force is less than) the effect of consciousness on behavior.

Assuming that this justification holds, we shall now examine the episode of **scratching the armchair-knees** in greater detail. First we must note that this episode, based mainly on the scratching response, which is the central behavioral element, and which is explained mentalistically, contains several more behavioral elements, some of which require a mechanistic explanation:

- In the late evening, when I am on my own, Max sees me sitting in my armchair watching television;
- Max comes up to me;
- Max scratches the left-hand edge of the armchair;
- I turn to him, pick him up, set him on my knees, and stroke him.

As stated, some of this series of behavioral elements, such as seeing me sitting and watching television, call for an explanation that appeals to cognitive processes analogous to the operations of the computer, that is, to a mechanistic means of explanation. By contrast, the central behavioral element in this episode, the core of the episode that generates the pleasant interaction between the cat and me, an interaction that realizes the cat's will, is an element that requires a mentalistic purposive explanation.

Secondly, to begin the three-stage interpretation of the given episode we must ask the following question: How does the scratching response fit in with this interpretation? The answer is that the scratching response has undergone a change from an adaptive-survival function to a new purpose, namely winning petting, and this new purpose is identical to the purpose of the explanation by will/belief.

The scratching response itself, as all cat-owners know very well, is an adaptive-survival response: the cat extends its claws by opening its paws, when it is on the hunt or in defense-aggression situations, to help replace the claws or to mark the scratched place with its scent (see Taylor, 1986; Morris, 1997). The change in functioning of this response, from an adaptive-survival function to the purpose of being petted, is made by a process of learning.

Max learned that scratching the armchair in which I sit does not result in an immediate scolding, but the opposite: the response yields reinforcements. Often the scratching makes me lift him up, set him on my knees, and pet him. This answer is supported by the following observations.

Observation 1: Max does not usually scratch Aviva's armchair while she is sitting in it and reading a book, watching television, or chatting with me, because she immediately responds by scolding: "Maaax, stop that". By contrast, when Max scratches my armchair (the left-hand edge of the seat) I am not quick to scold him, like Aviva. As a result, Max has learned to scratch my armchair and to avoid scratching Aviva's. The reason for this difference in our approaches to Max's behavior stems from my softness toward Max. (Sloppy and silly behavior, which resulted in the ruin of our furniture, according to Aviva.) I also let him scratch my shoes when I get home from work in the evening and I see in this behavior an expression of Max's affection for me (which causes Aviva to raise an irate eyebrow).

Observation 2: When Max wants petting from Aviva he does not scratch her armchair, but assumes his sphinx posture beside her right leg and fixes her with a long, unbroken stare, makes whining noises to draw her attention, and waits for her call, "Come on Max, come here beautiful cat". Sometimes, when he loses patience, he simply jumps onto her knees uninvited.

Observation 3: When Aviva is sitting in her armchair, reading or watching television, Max is somewhere around her. When Aviva goes off to her bedroom, Max usually scratches my armchair while I am watching television. This mostly happens after midnight. (Aviva is an early-riser type, while I am a late riser.) When Max scratches my armchair while I am still sitting in it, I interpret his behavior as a sign of wanting to be petted, and I lift him onto my knees. This habit of Max's becomes more emphatic because if Max scratches other armchairs I scold him at once: "Maaax, stop that immediately".

This interpretation of the episode of **scratching armchair-knees** may be summarized as an example of the three-stage interpretation in the following manner:

a) Determining the general framework of the explanation: the purposive explanation: *Max* wants to achieve *petting from me* and believes (by virtue of earlier learning, acquisition of knowledge) that *scratching* the armchair will get him *petting*; therefore Max *scratches* the armchair.

b) Explanation of the scratching response: This response in itself is explained by an appeal to a mechanistic explanation, that is, to the anatomical-physiological structure of the claws and to their evolutionary functions: hunting, defense-attack, replacing claws, and marking the furniture (in the Rakover apartment) with the cat's scent.

c) Integrating the scratching response with the purposive explanation: How does the scratching response, whose explanation according to (b) is mechanistic, fit into the framework of the purposive explanation? The integration is accomplished by means of a learning mechanism, that is, Max has learned that under certain conditions the scratching response brings about the realization of his purpose: being petted by me. And now, frequently, when the wish arises in Max to be petted by me he uses this acquired knowledge to achieve his goal. This goal (petting) is entirely different from the previous functions that were achieved by use of the claws: while the previous functions were connected to hunting, defense-attack, replacement of claws, and dispersing the cat's scent, the new aim is different: petting (stroking, tickling, and other expressions of affection).

This interpretation calls for three important clarifications:

1) *Function and purpose*: I use the term "purpose" as referring to part of the individual's internal conscious mental world, for example, when the individual has the intention of seeing a movie and she realizes it by going to the cinema. This term does not suit the explanation of the function of the action of the heart (to make the blood flow) or the explanation of the function of innate behavior

such as reflexes and instincts. The term function is suitable for these examples, making it possible to explain a given action from two points of view.

First, an action performed by the individual or by part of the system can be perceived from an adaptive and evolutionary angle. For example, if the action of the heart is flawed the individual's survival value declines, and if the claw-extension response in the cat is harmed its ability to survive and multiply is impaired. From this standpoint, it may be said that the explanation in the case of the **scratching the armchair-knees** episode is based on the change that has occurred in the function of the scratching response in Max: from an innate function, determined by evolutionary biological processes, to a conscious purpose, determined by the intention, the free will, of the cat.

Secondly, a given action can be perceived as a behavioral element that fulfills an appropriate function in the entire system. I shall expand on this matter later on. (Although a broad philosophical discussion of the term 'function' is beyond the design of this book, it is worth noting that in Searle's opinion (1992), attribution of a function to part of a system, or to a given behavior, is made from the viewpoint of the one who ascribes the given function. This criticism, it seems to me, takes us back to the discussion in chapter 1 on anthropomorphism, and my answer to this criticism, therefore, is similar to that which I offered in that chapter.)

2) *Explanation and change*: Why does the three-stage interpretation stress the change in the function of the central behavioral element? It is because an explanation is given for a change in behavior. Scratching the armchair in the context of the **scratching the armchair-knees** episode constitutes a behavioral change in respect of the cat's typical standard scratching response. This change is not explained as a change in the response itself, but as a change in the functioning of the response.

The epistemic need to give an explanation arises when a change occurs. For example, we ask, why has the light gone off? (Because the bulb has burnt out.) Why doesn't the car start in the morning? (Because the battery is dead.) We are not required to offer an explanation for an existing and understood situation: the light shines and the car starts in the morning, because in these cases the electric systems are functioning full well. (The answer to the question of what is considered a preceding and understood situation is of course relative. For example, a person may seek an explanation of how the car engine works.) And to the same degree, in keeping with Newton's first law, the law of inertia, we are not required to offer an explanation for a body that persists in uniform movement in a straight line unless a change has occurred in the direction or speed of the movement. The explanation for the present behavior being researched, then, is given in the setting of the behavior that preceded it, that is, in the setting of the difference between the researched behavior and the preceding behavior. (From the viewpoint of planning a research project, the present case fits a within-subject design, called pretest-posttest, when the behavior 'before' is compared with behavior 'after', i.e., after the independent variable has been operated; see, e.g., Neale & Liebert, 1986.)

3) *The Principle of New Application*: The three-stage interpretation has referred so far to the first condition of the Principle of New Application: the application of an existing response for the purpose of achieving a new goal, different from the previous function (adaptive-survival) of this response. How then does this interpretation relate to the second condition of the principle: application of different behaviors (including new responses) for the purpose of achieving the same goal? To suggest an answer we shall look at the following three examples. In the behavioral episode **rolling on the back: continuing to pet**, Max stands in front of me, rolls over onto his back, turns his white belly to me, fixes my eyes with his blue eyes, and I bend over him, sometimes to tickle him on the belly and under the jaw, and sometimes I set him on my knees. The interpretation of this behavior is similar to the **scratching the armchair-knees** episode, therefore, according to the second condition of the Principle of New Application, rolling on the back is an additional response to the armchair scratching response that results in the achievement of the same purpose: petting.

Max also achieves the same goal (eating and drinking in the present case) by performing a fairly new response pattern, which I never observed until its first occurrence in the episode of **two routes to food**: Max can get to the food and water located at the side of the kitchen by way of the lounge and through the kitchen door; or he can go through the bathroom corridor. Max uses both. One day I forgot to open the bathroom door that leads to the kitchen after taking a shower. (The doors of the kitchen, the bathroom, and the kitchen porch are always open.) While I was sitting wrapped in my bath robe in my armchair in the lounge, I saw Max leap over the chair on the porch and walk to the passage leading to the bathroom. After a few seconds he retraced his steps and went straight into the kitchen, traversed its entire length, reached the food, and began to eat. How may this behavior be interpreted? I believe that Max behaved as would a human who wishes to reach a restaurant and finds that the road is blocked (because of road repairs): that person would go back the way he came and reach the restaurant by an alternative route. Max's behavior then can be seen as an example of achieving the same purpose – eating and drinking – by means of a new response: {walking forward on route A (via the bathroom corridor), retracing his steps, and proceeding along route B (through the kitchen door)}. It is entirely reasonable to assume that Max used information acquired previously to achieve the same purpose (satisfying hunger and thirst) that he had achieved till then by taking either of these two routes.

5.2 Comparison of the three-stage interpretation and other approaches to an explanation for complex behavior

As may be imagined, many researchers have tried to develop different approaches to deal with complex behavior. I shall discuss two such approaches, which are based similarly to my approach to analysis of a complex behavior into its components: one is called "functional organization of behavior", connected to ethological research, and the other is called "functional analysis", connected to computer science, cognition, and research in engineering.

Functional organization of behavior: In a summary article Baerends (1976) analyzed the functional models prevalent in ethological research and showed that complex behavior of animals is subject to description and explanation by a hierarchical system containing a large number of "fixed action patterns", controlled by a smaller number of behavioral subsystems, which in turn are controlled by two or three general behavioral systems. This hierarchical approach makes it possible, among other things, to deal with the fact that instinctive action is rigid on the one hand (fixed action patterns) and flexible on the other – it has a purpose, it depends on internal and external factors, and it has a neurophysiological base. As an example of this approach let us study the system called "hierarchical organization" of Tinbergen (1951/1969), which is illustrated through the reproductive behavior of a fish called the three-spine stickleback.

In spring, with the lengthening of the days and the rise of water temperature, the male enters the first hierarchical level of the reproductive instinct, that is, a motivational center awakens in it that on the one hand is responsible for relevant behavior such as wandering (to find a territory to build a nest), and on the other hand this motivational center arouses more centers, on the second hierarchical level (i.e., subsystems): aggression, nest-building, mating, caring for offspring. Each of these centers in turn arouses additional behavioral centers on the third level; an example is aggression, including fixed response patterns such as pursuit, biting, threatening (these responses are called "consummatory acts": see Tinbergen, 1951/1969, p. 104). This instinctive response complex largely depends on environmental stimulation. For example, the aggressive response pattern depends on the appearance of another male fish invading the territory in which the nest has been built by the fish that possesses the territory. Which aggressive pattern the nest-building fish will use to respond depends on the invader's behavior: if the foreign invader bites, the territory-holder will bite; if it threatens, the territory-holder will also threaten; if it flees, the householder will chase it.

According to Tinbergen, then, instinctive behavior is intricate and complex, and is explained hierarchically (according to the given level) by interaction between central excitatory mechanisms (CEM) and external factors, environmental stimuli, called "releasing factors". These stimuli operate an internal mechanism, called the innate releasing mechanism (IRM), which removes a block that does not allow the CEM to disassemble and activate the characteristic behavioral pattern that is under the control of the center in question. The moment the appropriate stimulus appears in the environment the block is released by the IRM, and the individual responds by the behavioral pattern that is fixed and characteristic of its kind, and of the level of the center. In general, the degree of behavioral flexibility diminishes with the lowering of the CEMs (from level 1 to level 3).

This model of Tinbergen was able to explain several behaviors in animals. Among other things an explanation was given to a phenomenon called displacement activities, for example, the digging of pits by the three-spine stickleback precisely when this fish presents aggressive behavior toward another fish that invades its territory. The explanation is based on the fact that in states of conflict, tension, and high excitement, the motivation center, charged with great energy, dismantles itself by activating another center responsible for behavior that does not belong to the interactive behavior in which the fish was engaged at that time. (It is interesting to note that a similar behavior was observed in Max: **diverting aggression**. Max awaits with great anticipation being brushed by Aviva after she returns form her morning walk. Several times, when Aviva was preoccupied with other things, the cat's patience ran out, as he waited for her on the telephone stand, until he could bear it no more; then he pounced on her armchair and began scratching it. A similar phenomenon has been observed in a large number of animals, including humans: withholding of a positive reward causes frustration, which is released by diverting aggression to other stimuli in the surroundings.)

The three-stage interpretation and the hierarchical organization approach evince several similar features, such as separation of a behavior into its components, and a level of analysis that basically is not neurophysiological but functional. Still, my approach and the hierarchical organization approach fundamentally differ: while the latter is based on a mechanistic (biological) explanation, which deals with the complex of components of the animal's behavior in certain conditions (aggression, diverted responses), the three-stage interpretation attempts to interpret a behavior according to the mentalistic purposive explanation (will/belief). The core of the explanation lies in the functional change of the central behavioral component from an adaptive-survival function to a new function – the attainment of petting, a goal that well matches the explanation by will/belief.

Functional analysis: In his book *The nature of the psychological explanation* Cummins (1983) proposes an explanation for a complicated system through conducting a functional analysis, such as an explanation of the operation of assembly-line production:

> Production is broken down into a number of distinct and relatively simple (unskilled) tasks. The line has the capacity to produce the product in virtue of the fact that the units on the line have the capacity to perform one or more of these tasks, and in virtue of the fact that when tasks are performed in a certain organized way – according to a certain program – the finished product results. (p. 28)

Diagrams in electronics, computer programs, physiological analysis of an organism, are examples of this kind of functional analysis. Functional analysis, then, breaks a given system down into its simplest, most elementary components and shows how the organization of simple parts produces the complicated behavior of the entire system. For example, in the case of a computer program, breaking down the system into its components, and breaking these components down into their components, in the end leads to the most elementary components, to basic logical operations done on the symbols zero and one, which can be expressed physically as follows: zero is translated into a state of absence of electrical current and one is translated into a state of electrical current. The interpretation of this translation from the symbolic system to the electrical system is that the physical system realizes the software, or that the software is applied on the physical system (see also discussion in Rakover, 1990).

In this case too the basic difference between the three-stage interpretation and the explanation proposed by means of functional analysis is this: while the latter explanation is mechanistic, in the case of the three-stage interpretation integration of the components of the investigated behavior takes place in the framework of a mentalistic explanation. While the decomposition of the system according to Cummins's approach goes as far as the most elementary parts – the "behavioral atoms", in the case of the three-stage interpretation the discussion ends with the change of the behavior/purpose connection. While in Cummins's approach the entire functional system (the software) has its grounding in the physical system, in the three-stage interpretation there is no such grounding. In my view, these differences between the system explained mechanistically and the system explained by an appeal to mentalistic processes reduce in the end to the fact that the functional analysis will be hard pressed to deal with conscious behavior (intentions, desires, and knowledge) because of the body/mind problem (and see a wide-ranging discussion later, especially chapter 8). By contrast, the three-stage interpretation, from its very use of mentalistic explanatory concepts, will not find it difficult to handle conscious behavior, that is, a change in behavior of this kind – a change in the behavior/purpose connection.

5.3 Cannot Max's behavior, ultimately, be explained mechanistically, as simple learning?

The three-stage interpretation offered an efficient account for some of Max's everyday behavior. This interpretation is based on a learning process whereby the central behavioral component, for example, the scratching response, has acquired a new purpose (different from its typical standard function, namely the adaptive-survival function) which fitted well into the framework of the mentalistic purposive explanation (will/belief). In light of the importance of the learning process in this explanation the following question arises: cannot the **scratching the armchair-knees** episode be explained as a simple learning process? Cannot the mentalistic purposive explanation be given up? To my mind the answer is no, because this episode is endowed with several qualities that are not characteristic of a learning process and that can be explained by an appeal to the mentalistic explanation.

The discussion on this question will be divided into three sub-sections: first I shall present several behavioral instances whose interpretation by means of a learning process is direct and clear; second, I shall discuss whether learning is explained by a mentalistic or a mechanistic process; finally I shall probe whether learning-mechanistic explanations may be suggested for mentalistic behavioral episodes.

5.3.1 Examples of behavioral episodes explained as simple learning processes

The phone ringing: When Max is sitting on Aviva's knees or on mine and the ringing of the phone is heard, he jumps down even before we get up to answer it. This, in my view, is simple avoidance learning that the cat learned from the dozens of cases in which the following sequence of events recurred: phone rings – we quickly rise – Max is dropped from/jumps off the knees – we go to the phone. If we assume that the event of "rising" is not pleasant for Max because rising causes him to be dropped/impairs his balance, we may regard the phone ringing as a warning signal and his response of jumping off the knees as avoidance behavior – a response that forestalls and prevents the occurrence of the unpleasant rising/being dropped/jumping off event.

Cleaning the apartment: When Aviva sweeps and washes the floor of the room, Max jumps onto the sofa from where he watches her actions. At first Max would chase after the broom or floor cloth and Aviva scolded him, until he learned to avoid this. This episode too may be interpreted as avoidance learning. The broom, the pail of water, the floor cloth, and the cleaning process itself are a kind of warning signal: Max, sit quietly on the sofa and don't get in the way.

Arranging the cover on the sofa: At first when Max lay on the sofa and Aviva straightened the cloth cover spread on it, Max would jump off the sofa, and jump back on after the cloth had been smoothed. In time Max learned to stay on the sofa while Aviva arranged the cloth. I believe that this learning is nothing but a process of adaptation. Contrary to the **phone ringing** episode, in which Max was dropped from/ jumped off the knees, in the present case Max has learned that the flapping of the coverlet under him does not entail any unpleasant event at all.

Displays of affection and satisfaction: When we are petting Max, the cat sometimes expresses affection for Aviva or for me by gentle biting of the hand. He lightly grasps the hand between his two forepaws without extending his claws and bites without his teeth piercing the skin. These two responses are voluntary and under his control. Another way in which Max shows affection and satisfaction is purring: when he sits on our knees and we stroke him he begins to purr. The latter response, according to Morris (1997), indicates friendship and is connected to the suckling time of kittens when purring tells the mother cat that everything is in order. These two responses may be interpreted as a learning process of generalization of stimuli: Max generalizes the same response, the slight biting, which appears in kittens' games, to us (Aviva and me); and Max also informs us, by purring, of his satisfaction. (Sometimes the purring response is manifested when we clean his eyes too.)

On the assumption that these examples indeed express learning processes, the question arises of how to explain them: by mechanistic or mentalistic means?

5.3.2 Are learning processes mechanistic or mentalistic?

One of the important aims of this book is to suggests behavioral guidelines (see above, the Principle of New Application; the Criterion of Mentalistic Behavior) by means of which it will be possible to discern when it is apt to propose a mentalistic hypothesis, that is, a hypothesis that posits a mentalistic explanation for a given behavior or behavioral component. In other words, I am looking for a yardstick to distinguish mechanistic from mentalistic behavior.

This distinction risks being confused with the well known distinction between innate behavior and learnt behavior. To dispel any lack of clarity, I wish to stress here that there is no parallel between the mechanistic/mentalistic explanation distinction and the innate/learnt distinction. As we shall see below, in most cases innate and learnt behavior have been given a mechanistic explanation, even though some of the learnt behavior allows a mentalistic explanation. To support this, in the present discussion I shall concentrate, as succinctly as possible, on some of the work of Lorenz (1950, 1965) and of Tinbergen (1951/1969), the founders of modern ethology, who developed the important concepts of instinct, innate/learnt behavior, and I shall show that their explanation is mechanistic. Furthermore, I shall show that the discussion and the debate on the innate/learnt dichotomy does not parallel the mechanistic/mentalistic explanation dichotomy, because the innate/learnt dichotomy takes place entirely in the framework of the application of the mechanistic explanation model, and its purpose is to test the degree of efficiency of various mechanistic theories.

In chapter 4 I proposed a behavioral criterion called the Criterion of Mechanistic Behavior: *If an* ***animal of a given species*** *is in a state of a* ***certain physiological drive*** *and in an environment of* ***stimuli relevant to that drive****, then a* ***characteristic and fixed pattern of responses*** *will arise in it.* I illustrated this through analyzing several behavioral episodes of Max the cat. Although a behavior that satisfies this criterion seems to call for a mechanistic explanation (neurophysiological, genetic, evolutionary), this criterion should not be regarded, as stated above, as anything but an indication, a clue that suggests or invites a *mechanistic hypothesis* (a hypothesis that posits a mechanistic explanation). Why? Because this criterion is liable to be too broad; for example, it is easy to make the mistake of thinking that rigid behavior habits are innate and may be subject to a mechanistic explanation. It is said that the daily routine of the revered philosopher Immanuel Kant was so meticulous that people could set their watches by the moment the philosopher left his house for his afternoon walk. Yet although the philosopher's behavior shows a characteristic and fixed pattern of responses that appears in the presence of suitable states of stimuli and drives, we would not tend to suggest a mechanistic explanation for his behavior, simply because this behavior, this daily routine, of the philosopher was determined according to Kant's free will.

Moreover, it is hard to find instinctive behavior that is not influenced by environmental factors and learning processes. Tinbergen (1951/1969) himself writes in his preface to a reprinting (in 1969) of his famous book of 1951 *The study of instinct*:

> … a rigid dichotomy between 'innate behaviour' and 'learnt behaviour' is no more than a first hesitant step in the analysis of the developmental process as a whole… (p. viii)

Still, Tinbergen maintained that at root his approach was correct, and even suggested that the effect of learning on the development of instinctive behavior –what is learnt and what is not learnt – this too is determined by innate, programmed, factors. This approach is mechanistic in essence.

As is seen from the example of the "hierarchical organization" analysis of Tinbergen (1951/1969), the reproductive behavior of the three-spine stickleback above, the model that emerged, despite its complexity, is mechanistic in quality and may be characterized as a mechanism that acts according to several relatively simple rules of interaction between internal factors (motivational centers that are filled with suitable instinctive "energy") and external, environmental stimuli, which release the energy stored in these centers and as a result produce the characteristic instinctive behavior. (Tinbergen's model is very similar conceptually to the motivational-hydraulic model proposed by Lorenz (1950): in this model, based on an analogy to a hydraulic system, a valve, representing the IRM, is opened by a weight, which represents the weight of the environmental stimulus perceived by the animal, and as a result of this opening a quantity of energy is released from the motivation-energy reservoir, which activates the appropriate behavioral pattern.)

Furthermore, the hierarchical organization method accords with the four aims, questions, of ethological research, as Tinbergen (1963) stated in his renowned paper (which I repeat here): (1) to discover the immediate causes responsible for a given behavior; (2) to discover the function of the behavior (adaptation, reproduction, survival); (3) to discover how the behavior develops during the lifetime of the animal; (4) to discover the evolutionary source of the behavior, the link with ancient generations. These aims, in Tinbergen's view, complement each other. As I argued in the last chapter, it is hard to find in these aims any connection to questions about the animal's mental functions, its feelings, thoughts, intentions, desires, and states of consciousness (e.g., Jensen, 2002). That is, ethological research follows the trail blazed by mechanistic explanations.

The model of Tinbergen (1951/1969) and his classic ethological approach, as well as that of Lorenz (1950, 1965), to instinctive, innate/learnt, behavior, drew severe criticism, such as the mismatch between the energy-motivations concept and the actual nervous mechanism that activates the animal's movement; overemphasis on internal factors, without consideration of sensory feedback; restriction of the function of environmental stimulation to releasing the motivational energy; and drawing dubious conclusions about the innate/learnt dichotomy from 'deprivation' experiments. (See summary and discussion of these matters in Dawkins, 1995; Eibl-Eibesfeld, 1975; Lehrman, 1970; Lorenz, 1965; Kennedy, 1992.) Here it is worth expanding a little on Lehrman's critique and on the debate that took place between him and Lorenz.

In 1970 Lehrman published a critique of Lorenz's innate/learnt dichotomy; it was based on an earlier paper (Lehrman, 1953) criticizing Lorenz's concept of instinct. In the 1970 paper Lehrman also took issue with Lorenz's (1965) response to the criticism leveled against him. I wish to highlight the following points in that article, which seem to me especially relevant in a discussion of the innate/learnt dichotomy. In Lorenz's view, the biologist's function is to understand how the genetic pattern, which is created by a process of natural selection, arouses normal behavior in a normal environment. This has to be tested by a deprivation experiment, in which one tests whether the studied pattern of behavior will also develop in a situation where the animal is deprived of its normal stimulation, for example, the experimental design where the animal is reared in a certain social or sensory isolation. If despite this deprivation of the relevant stimuli the behavior pattern develops as in animals not reared in such circumstances, we can conclude that this behavior pattern is innate, and controlled by innate factors. Lehrman criticized this approach from several standpoints: first, he argued that the structure of the behavior depends on the internal development of the individual as well as on the interaction between the innate tendency and the appropriate environment that allows development of the given behavior pattern. A behavior is not detached from the environment and can develop only in the right one. Secondly, he points out that conceptually the geneticist's grasp of the notion of innateness (connected with the distinction *between* subjects in the same environment, that is, if in the same environment different individuals have different behavior patterns, the explanation is by an appeal to genetic factors) is different from the concept of innateness as Lorenz perceived it (connected with fixed development *within* the same individual, that is, behavioral development of the same individual, which is not influenced by learning processes). Thirdly, he argues that methodologically the inference concerning innate behavior as a result of lack of effect of certain stimuli on the studied behavior is extremely weak, because what the experimenter seeks is not reasons for non-change in behavior but reasons on account of which the behavior changes and is created. And fourthly, he stresses that by his approach Lorenz neglects the earliest stages in development, and the fact that the environmental effect operates not only through learning processes but also as a result of other changes, such as temperature and nutrition.

Clearly, this critique of Lorenz's approach does not turn on the question of which kind of explanation should be chosen in order to explain a given behavior mechanistically or mentalistically (the critique does not mention behavior linked to the animal's mental-conscious states), but about the limitation of the mechanistic model, which treats behavior as innate or as learnt, and the experimental methodology on which this model rests.

To exemplify how far a behavior which in essence is innate and characteristic to a species depends on learning in its development, let us peruse once more hunting and preying behavior in the cat (see Morris,1986; Bradshaw, 2002). Even though the components of hunting behavior in the cat are innate and controlled by the hypothalamus, it transpires that perfected preying behavior is not achieved without the mother's coaching. There is great importance in the fact that the mother cat brings prey to her kittens and that these learn to pursue the quarry and bite it in the nape. These behaviors are learnt by observation of the mother's actions and of the others in the litter. Morris describes research findings according to which out of twenty kittens separated from the mother (in a Lorenz-type deprivation experiment), only nine killed prey spontaneously, and of these only three ate the prey. Kittens that have not received training in hunting between the sixth and the twentieth week of their lives will not become efficient hunters. Moreover, for kittens that grew up in the company of rodents, the rodents ceased to be prey. That is, even though the behavior pattern of hunting-killing is innate, environmental conditions are likely to teach the cat a behavior contrary to its nature.

The picture that emerges from the present discussion is this: behavior is a function of a complicated interaction between genetic factors and environmental factors, where the latter are subject to division into learning and non-learning factors (e.g., temperature, nutrition). (As may be realized, the story is far more intricate. For example, Lehner (1996), in his book on ethological methodology, suggests a behavioral model whereby the tendency to behave is a function of the following factors: the genotypical, the environmental, the interaction between these factors, the anatomy of the animal, and its physiology.)

In light of this discussion, it may be suggested that the probability of a mechanistic explanation to handle non-learning genetic influences and environmental influences is very great. Can a mechanistic explanation also be applied to learnt behavior? The answer is not simple. Much of what a human being learns is through a conscious process of acquiring knowledge (e.g., studying at school, at university) so it seems that this learning calls for a mentalistic explanation. Still, it turns out that the explanations given to learning by animals are mechanistic, and are based on three main mechanisms: associative mechanisms between stimulus and response (including models based on hypothetical mechanisms or intervening variables); cognitive mechanisms based on analogy with computer software; and biological, neurophysiological, mechanisms. Furthermore, it appears that many of the explanations for the acquisition of knowledge in human beings are also based on a cognitive mechanism analogous to the computer (see a wide-ranging discussion on these and other subjects in Benjafield, 1997; Domjan, 1998).

Now, in light of this discussion, let us study the examples in the last section and ask: can a mechanistic or a mentalistic interpretation be proposed for them? My argument is that these learnings are explained by an appeal to mechanistic processes. As an illustration we may examine one behavioral episode, that of the **phone ringing**. The explanation I offered for this learning was based on avoidance learning (the explanation for this learning has undergone theoretical-empirical changes that I cannot discuss here. See reviews in Domjan, 1998; Makintosh, 1974). One of the important theories, the pillar of avoidance learning, is the two-stage theory of Mowrer (1947): at the first stage, an association is established through classical conditioning between a stimulus (light) and punishment (electric shock), so that the appearance of the light arouses a fear response in the individual; in the second stage the individual learns to respond in the presence of light with a response that causes the fear to cease (by stopping the light), which constitutes reinforcement of this response. (Reinforcement is a stimulus that raises the probability of the appearance of the response to which this stimulus is connected, for example, pressing on a lever which yields a grain of food.) As may be seen, the dual theory explains how the individual has learned to avoid punishment by an account of a mechanistic means: the light arouses the response of fear, which arouses a response that stops the light – fear. This explanation is applicable to the present episode. The ringing of the phone arouses in the animal a negative experience (anxiety or fear, connected with the cat's experience of losing its balance by being dropped), which arouses a response of jumping off the knees, which prevents the unpleasant experience, the experience of fear: Max jumps off the knees immediately on hearing the ringing of the phone.

5.3.3 An attempt to propose mechanistic explanations for mentalistic behavioral episodes

As a result of the present discussion, that learning in animals is explained by an appeal to mechanistic explanations, the need arises for extra caution in applying the mentalistic explanation to a given behavior. We must therefore study whether despite all it is not possible to propose a mechanistic explanation for behavioral episodes to which we have applied the three-stage interpretation.

As the first example, we shall again study the **scratching armchair-knees** episode: Max wants me to pet him; he approaches the armchair in which I am sitting and begins to scratch the upholstery with his claws. I bend down to him, set him on my knees, and stroke him. Cannot this episode be interpreted as simple instrumental learning of the state of stimulus–response–reinforcement kind, where as a result of it, the frequency of the appearance of the response in this state of stimulus increases? The answer is negative, not because this episode is not based on a learning process that is likely to have a mechanistic explanation, but mainly for the following reasons.

This behavior is not stereotypical, recurring in the same situation. Max does not scratch the armchair every night, or whenever an appropriate state of stimulus is created, that is, a situation in which I am watching television alone at night; he continues to scratch all the other chairs and armchairs in the room; he jumps onto my knees (from the left) also without scratching the armchair beforehand (see the **knees-beating** episode, below). This episode therefore cannot be treated as a simple case of learning, in which a given stimulus produces a learnt response pattern over and over again. Max is the one who decides when and how to realize his intentions; it is he who initiates this episode, the interaction between him and me. It is reasonable to suppose that Max uses this acquired information (scratching – lifting up the cat and placing him on the knees – petting) when the desire arises in him to be petted by me. And just as we generally go to a restaurant to allay hunger pangs, and use acquired knowledge (e.g., the location of the restaurant, the kinds of food served) to satisfy our hunger, so it is with Max, who customarily scratches the seat of my armchair when the desire arises in him to be petted. In other words, the scratching response serves the cat's wish and belief. I think that Max uses his ability (to scratch, etc.) in order to achieve his goals similarly to our use of tools to realize our goals. For example, we use a flashlight to illuminate. And although it is very easy to suggest a mechanistic explanation for the question, how does the phenomenon of illumination of the flashlight take place? (by the theory of electricity), it is not possible to use this mechanistic theory to answer the question, why have human beings invented this tool? By the same token, one may suggest mechanistic explanations for Max's motor movements and for learning processes, but these explanations will not be right for the question, why does Max decide, from time to time, to scratch the armchair when I am sitting in it and staring at the television?

A further possibility for a simple learning explanation is this: Max has learned that he may scratch my armchair without my scolding him, therefore, he scratches the left side of my armchair seat – and that is all that Max does! The rest of the episode is explained as follows: *it is I who think that Max wants to be petted (when all Max wants is to perform the scratching action) and it is I who lift him onto my knees and pet him.* Is this interpretation reasonable? I don't believe it is, because it does not accord with the following observations: first, when we initiate the setting of Max on our knees without his seeking it, he tries to get off and run away. By contrast, not once has Max tried to get off my knees in the **scratching armchair-knees** episode.

Secondly, occasionally, when I am sitting glued to the television with my left arm lying on the armrest of the armchair, Max taps my arm with his paw; and when I turn my head to him he sometimes begins to scratch the armchair, sometimes he starts climbing onto the armchair, and sometimes he sits there waiting for me lift him onto my knees (**tapping the hand**). So its is reasonable to assume that in this case, as in the foregoing, it is Max who has initiated the interaction between him and me, once by scratching the armchair and once by tapping me on the arm.

Thirdly, in most cases, when Max wants to beat me on the stomach he jumps directly onto the left armrest of the armchair, and from there onto my knees, and while seated on his rear he beats my stomach. When I try to stop the beating and set him on my knees he jumps up and gets off my knees (**knees-beating**). This is a different behavior from that of the **scratching armchair-knees** episode and that of the **tapping the hand** episode. Presumably, then, these behaviors (scratching the armchair seat and tapping the hand, as against jumping directly onto the armrest) have different purposes (petting as against beating the stomach).

As a further example we shall take another look at the episode of **rolling onto the back: continuing to pet**. Max stands in front of me, and rolls onto his back, turns his white belly up to me, and fixes me with the blue eyes. I bend over him, and sometimes I tickle him on the belly and under his chin, sometimes place him on my knees. Cannot this behavior be interpreted as based on an innate play behavior in cats?

Bradshaw (2002) states that the "belly-up" posture (which I called "rolling onto the back") is an important and common response that appears in social play in kittens. The play involves pairs that take turns at response counter-response; for example, cat A charges and pounces on cat B, which responds with belly-up, cat A: belly-up/cat B: standing, and so on, where the three responses – charging, belly-up, standing – appear the most frequently in relation to other play responses, such as pursuit. That is, to distinguish play from a real fight the cat shows its partner a response that signifies its intention: a fight-like game, and not a real fight (just as children play friendly war games). (Postures that signify intention of play appear in other animals too. For example, Bekoff reports that in dogs, wolves, and coyotes the "play bow" posture – forelegs stretched out forward and rear legs stiffly upright – means play and not a fight. See description and discussion in Allen & Bekoff, 1997.) So cannot the present episode be regarded as an expression of social play among cats, where I am perceived in Max's eyes as a part of the family (a large cat?)? The answer is no. The negation does not disqualify the meaning of "rolling onto the back" as a response signaling peaceful intent, and not a fight, but emphasizes the additional meaning that Max has inserted into the response, that is, the use he makes of this response to achieve other purposes: getting attention, petting – situations in which Max is not active, as in play, but passive, receiving positive reinforcements from me.

By the nature of things, the games played by me and Max showed no ordered response and counter-response that characterize kittens' play. Still, the elements of play may be observed. We shall look at the **rolling onto the back – contacts** episode: Max sprawls on his belly in front of my armchair. I bend over to him and let the fingers of my right hand come into contact with his head or his rear legs. Max rolls over onto his back and we begin the game: Max tries to grab my right hand between his fast forepaws and I try to escape them and touch his body. The game ends when I sustain a scratch, or when Max leaves the arena because he cannot defend himself when I bring my left hand into play.

Although in this episode the meaning of the "rolling onto the back" posture is very similar to this response in kittens' play, this meaning changes in the **jumping and running–rolling onto the back** episode. In this case, while I am walking Max approaches me from behind at a run, leaps beside me (usually on the right side) while making punching gestures with his forepaws, runs in front of me, rolls onto his back, and blocks my path. So far, the account seems like a partial description of the chain of responses that characterize play in kittens (pouncing, belly-up). But the similarity ends there, because the rest of the episode is not play as in the foregoing episode, but petting. When I bend over Max he does not play with me actively but abandons himself wholly to a series of stroking and tickling on the belly, neck, chin, head, and behind the ears. So in Max the "rolling onto the back" posture has acquired new meaning, which invites from me (and from Aviva) an attitude expressed in petting. This significance also finds expression in the following episodes (which I discussed earlier): **armchair game–petting, sleep: running–rolling onto the back**, in which rolling onto the back succeeded in winning from me a social reward, an attitude of affection, and petting.

This interpretation, that Max uses the "rolling onto the back" response to achieve a new purpose (petting, and not play in which two kittens are actively involved in actions that imitate a fight), raises a further question: may not a simple learning explanation be offered for the fact of achieving the new purpose by means of an existing response? For example, let us assume that Max rolls onto his back with the aim of playing a kitten-game with me, *but it is I who decide to pet him (stroking, tickling belly, neck, chin, head, behind the ears)* and as a result Max has simply learned that "rolling onto the back" incurs petting. (That is, like the interpretation above, here too an interpretation may be suggested for the given response by an appeal to simple instrumental learning of the stimulus–response–reinforcement kind, in which the association between a stimulus and a response is strengthened.) If indeed this interpretation holds, it emerges that it will be hard to interpret this behavior by appeal to a mentalistic explanation. My answer to this question is negative.

Even if we assume that possibility (a): Max has learned the "rolling onto the back"–petting connection through simple instrumental learning seems more logical or realistic than possibility (b): Max initiated communication with me, and wanted petting from me by "rolling onto the back", we will still be hard pressed to interpret the following observations according to possibility (a).

First, similar to what was stated above, this behavior (rolling onto the back) is not stereotypical, recurring in the same state of stimulus. Max does not roll over onto his back every time he is in front of me (or Aviva), that is, although the stimulus is appropriate (e.g., I am sitting in the armchair) Max does not respond immediately with a response connected to this stimulus. In most cases, when Max is in front of me, or Aviva, whether sitting or lying, he does not roll over onto his back, but simply goes on with this behavior, and afterwards he goes to other places in the apartment. That is, Max initiates and decides how to realize his intentions.

Secondly, Max uses the "rolling onto the back" response in several cases to achieve petting and a companionable attitude from me: in the **jumping and running–rolling onto the back** episode, Max passed me, rolled over onto his back, and begged for petting from me; in the **armchair game–petting** episode Max rolled onto his back and begged for petting after he jumped onto my armchair when I was just about to sit down in it; in the **sleep: running–rolling onto the back** episode Max passed me at a run before I left for my bedroom, and rolled over onto his back (sometimes he simply waited for me on the bedroom threshold) begging for petting and my company (Max learned that when I turn off the TV and the light I am going to my room, and he is left alone). It is hard to explain these cases (as well as the **rolling onto the back: continuing petting** episode) as based on simple learning of response (rolling onto the back)–positive reinforcement (petting). Possibility (b) is more likely, that is, after Max learned one way or another the connection between rolling onto the back and petting, he used this knowledge in other situations also to convey his intentions to me. It is Max who initiates the communication between him and me.

More evidence supporting possibility (b) is the fact that Max achieved petting not only by rolling onto the back; as we saw above, he achieved the same aim – petting – both by scratching the seat of the armchair and by jumping onto the knees.

Third, as stated above, rolling onto the back is one of the important and characteristic responses to fighting play in kittens. Therefore, it is reasonable to suggest that the rolling onto the back–play connection is deeper rooted and stronger than the learnt rolling onto the back–petting connection. On this assumption, we would expect that when Max rolls onto the back a series of responses typical of play will appear at greater frequency than complete surrender to stroking, tickling, and light squeezing of the belly. However, what actually happens is precisely the opposite. I found that of all the cases in which Max rolled onto his back, the number of cases in which this posture led to active kitten-like fight play, in which Max tried to grab my hand with this claws, to bite it (not fiercely), and to make a scratching movement with his rear legs, was very small.

This analysis substantiates the fact that it is hard to propose a satisfactory explanation for the two episodes **scratching armchair-knees** and **rolling onto the back: continuing petting** by an appeal to a mechanistic explanation (a simple learning process or innate responses typical of cats). To understand these behaviors we must use the three-stage interpretation, based on the cat's intentions (his desire, his purpose) – mental processes expressed in the performance of an existing response to achieve new purposes, or of different responses to achieve the same purpose.

But this appeal to the private world of the cat is liable to prove a two-edged sword. The moment we have decided to appeal to the cat's internal-mental world, a methodological door has opened through which many and varied hypotheses are likely to march that will be based on the attribution of human mental processes with the goal of explaining the given behavioral episode. For example, we may examine the following explanation for the **scratching armchair-knees** episode: Max the cat scratches the seat of the armchair because he is sorry for me and wants to save me from the awful ennui that the television brings down on me. Although this amusing hypothesis (an appeal to the cat's pity, empathy, and grasp of television programs) seems at first glance far-fetched and distorted, I don't think it should be taken lightly, but as an alternative hypothesis to be placed before the scientific test. In the history of science, seemingly absolutely unacceptable hypotheses have more than once proved correct. Without going into a theoretical discussion of the degree of rational foundation of this hypothesis, it may at once be said that this hypothesis does not meet the empirical test: I have not found that Max's scratches on the armchair seat had any connection with the mood I was in (bored, not bored) or the kind of TV program that was on (interesting, not interesting). Therefore, the hypothesis is rejected.

Figure 1. Looking into the mirror

Figure 2. Looking at himself

Figure 3. On the piano

Figure 4. Head on the pillow

Figure 5. Head on the table-leg

Figure 6. Max and the Siamese cat (a)

Figure 7. Max and the Siamese cat (b)

Figure 8. Max and the Siamese cat (c)

CHAPTER 6

Multi-explanation theory

In this chapter a "multi-explanation theory" is proposed, intended to address complex behaviors whose behavioral components require several mentalistic and mechanistic explanations. In the natural sciences the *explanation model* uses a number of laws or theories to propose an explanation for a given phenomenon; however, the relation of *use* is the opposite in psychology: there the *theory* uses a number of explanation models divided into two kinds: mechanistic or mentalistic. The test of the multi-explanation theory may raise three problems that are liable to impair its functionality: its providing an ad hoc explanation, its lack of internal consistency, and its incomparability to other theories. In this chapter I offer procedural guidelines to resolve these three problems. Nevertheless, methodological-philosophical analysis shows that while the multi-explanation theory can be put to the empirical test like any theory in the natural sciences, an important difference exists between the way a theory in the natural sciences proposes an explanation and the way the present theory works. This difference lies in the fact that the multi-explanation theory is based on two kinds of explanation models, mechanistic and mentalistic.

An item is explained when it is placed in a large theoretical framework, but when I examine it under the electron microscope the item looks bigger than the framework.

In the last chapter I developed the three-stage interpretation with the aim of tackling Max's relatively simple behavior. The purpose of this chapter is to expand this approach and to develop it into a general one, a general explanation scheme, which treats more complex behaviors, those that can be decomposed into a network of behavioral components that are given to different mentalistic or mechanistic explanations. An example is the dualist theory of memory (see below). I call this approach the 'multi-explanation theory'. In this chapter I shall concentrate on the methodological-philosophical problems that arise as a result of using this theory.

6.1 An explanation model, an empirical test, and a multi-explanation theory

The question what is a scientific explanation (and the close connection between explanation and causation) has several interesting answers in the professional literature (see discussions and reviews in Hempel, 1965; Pitt, 1988; Psillos, 2002; Rakover, 1990, 1997; Ruben, 1993; Salmon, 1989; Woodward, 2002, 2003). Within the scope of this book I cannot summarize and discuss these matters. It is worth emphasizing the difference between the natural sciences and psychology in the explanation model/theory relationship; I shall concentrate on the infrastructure of the philosophy of the explanation, that is, on the explanation model that Hempel proposed, which in fact is an abstraction of the way in which explanations are given in Newtonian physics.

Let us look once more at Galileo's law. Imagine that Ruth visits the science museum, sees a simple illustration of free falling, and asks, "Why did all the bodies in the experiment fall 4.9 meters in one second?" The answer to the question will be: all bodies behave in accordance with Galileo's natural law of free fall: $d=1/2gt^2$. If you set t as one second in this equation, you will find that all the bodies (regardless of their weight and size) will fall d=4.9m.

This specific explanation serves in the general procedure that has the structure of a logical deduction (according to Hempel, 1965):

Assumptions: 1) Theory, natural law (e.g., $d=1/2gt^2$)
2) Particular conditions (e.g., t=1)

Conclusion: Prediction, description of the discussed phenomenon (e.g., d=4.9m).

Explanation: If the prediction and the observation are in accord, the phenomenon is explained by means of Galileo's law; if not, the phenomenon is not explained.

Hempel suggests, therefore, a scheme of explanation, a model (called the Deductive-Nomological (D-N) model), whereby every scientific explanation has the same explanatory structure. The assumptions must include: at least one true, empirical law of nature (i.e., a law well anchored to empirical findings and to theoretical arguments); particular conditions (which in psychology are also called the 'independent variables'); and the conclusion must include: a prediction, a description of the phenomenon derived from the assumptions deductively (logically, mathematically) (see Hempel 1965; Hempel & Oppenheim, 1948). And now, if the prediction, the description derived from the assumptions of the explanation, accord with the observed phenomenon, we may say that the phenomenon has obtained an explanation by means of the law that appears in the assumptions of the explanation, that is, we may say that this phenomenon was expected (on the assumption that the natural law is true) and that all bodies falling in a free fall to the ground behave thus.

Without entering into a discussion of the critiques of this model, and of alternative models proposed for it over time, it is important for me to stress the two following points:

First, it is possible to introduce various scientific laws and theories into the Hempelian model.

Second, if the prediction does not accord with the observed phenomenon, we may say that the law or the theory is not capable of explaining the phenomenon, and they are refuted. *But we may not say that the explanation model itself is not efficient*. These points require clarification.

The first point: the Hempelian explanation model is a scheme that describes how explanations should be proposed in all areas of science. All one has to do is to set in its premises different laws, for example, laws of the motion of bodies, laws in electricity and electromagnetism, and to operate on these empirical laws Hempel's explanation scheme in order to obtain a good prediction of the given phenomenon, that is, to obtain its explanation. Hence, that explanatory scheme, that explanation model, constitutes an explanatory repository for different theories, for different laws. I shall call this property *One explanatory model – many laws and theories*.

The second point: This quality – one explanatory scheme for many laws and theories – is what underlies the argument that observations do not refute the explanation model itself, because with the same explanatory scheme we shall say in one instance of law A that it explains the phenomenon under consideration well (the prediction accords with the observation), and in a second instance we shall say of law B that it does not explain the results (the prediction does not accord with the results). Similar things may be said about the method of the scientific test.

The scientific method for testing a theory is similar in structure to the Hempelian explanation model, where the chief difference between the two is this: while in the explanation we assume that the theory is true (or close to the truth), in the test of the theory we examine the truth of the theory. The explanation demands that the theory be true, that it is not possible to propose an explanation from a false theory. By contrast, testing the theory demands that the question of the truth of the theory be open, because if the theory is known to be true or false there is no need to test it. The classic method of testing a theory or scientific empirical law is the Hypothetico-Deductive (H-D) method (e.g., Glymour, 1980; Poletiek, 2001; Salmon, 1967):

> **Assumptions:** 1) Theory, natural law (e.g., $d=1/2gt^2$)
> 2) Particular conditions (e.g., $t=1$)
>
> **Conclusion:** Prediction, description of the discussed phenomenon (e.g., $d=4.9m$)
>
> **Test:** If the prediction and the observation are in accord, the specific theory is supported; if not, the theory is refuted.

This method, then, tests the degree of accord of the theory, the law, with reality by a logical derivation of a certain prediction and a test of the match between the prediction and the observation. If there is a match, the theory is supported by the observation; if not, the theory is refuted, is found to be incorrect. (The methodological analysis of this method raises several interesting problems that we cannot deal with here; see, e.g., discussion in Poletiek, 2001; Rakover, 1990.)

As may be seen, the method of empirical testing, the H-D method, like the Hempelian explanation model, is a method that treats a large number of theories and laws, so it too maintains a property akin to the scheme of one explanation for many cases, that is, *one testing method – many laws and theories.*

Now as we are speaking about one model, or one testing procedure, that treats a large number of theories, when the results of the treatment are once positive – the prediction matches the observation – and once negative – the prediction does not match the observation, the following conclusion arises. If the observations do not match what results from the theory, then what is refuted is the theory (which has not succeeded in furnishing an explanation for the given phenomenon) and not the method of empirical testing. The explanation scheme and the testing method, then, are indifferent to the results of the explanation and the test; they only guide us as to how to act, and we test by means of these suppositions whether indeed our theory gives a good explanation, whether the theory has been supported or refuted. And example will explain this matter well.

Assume that an amateur researcher in Galileo's day proposed a different law for the falling of bodies: $d' = 1/2gt^3$. This law explains very well the fall of the body in the first second. But what happens after two seconds? According to Galileo's law the body will fall: $d=19.6m$, and according to the new law: $d'=39.2m$. Obviously, the new law is refuted and Galileo's law obtains empirical support; the new law is not able to explain the results while Galileo's law does explain them. May we say that the refuting of the new law also refutes the method of empirical testing and the explanation model? Clearly, the answer is negative. If it were affirmative, it would not be possible to put any theory to an empirical test, because in principle one negative result, the lack of a single match between the predicted and the observed, would be enough to refute both the testing method and the specific theory under discussion.

And what happens in psychology? What happens when the structure of a theory in psychology is a multi-explanation structure? I believe that it is not possible to apply directly to psychology the property of *one method (explanation model, empirical test) – many theories.* To my mind, the situation in psychology is the reverse: here the same theory uses many explanation models, a situation that I call *one theory – many explanation models.* The question, of course, is why do I believe that in psychology the situation is the reverse? The answer lies in the fact that in psychology the explanation for a behavioral phenomenon is usually given by means of two kinds of explanation that do not match: the mechanistic explanation and the mentalistic explanation. This answer calls for a separate discussion, which I shall undertake in the next chapter. However, it is impossible not to allude to the theoretical path I shall follow.

One of the arguments, which I call the "explanation model argument", against the approach of "one theory – many explanation models" is the following. One of the important mentalistic explanations is the purposive explanation. For example, David drove in his car to Jerusalem because he wanted to visit his girlfriend, and because he believed that the journey by car would fulfill his wish. This specific explanation may be generalized inductively (beyond people, desires and beliefs) and the following *teleological law* may be proposed:

If X desires G and believes that B will realize G, then X will do B.

Now the explanation model argument proposes that this law can be introduced into the explanatory scheme of the D-N model in the following way:

Assumptions:	1) Law, theory: the teleological law 2) Particular conditions: David wants to meet his girlfriend in Jerusalem; David believes that a ride in his car will fulfill his wish
Conclusion:	Prediction: David will drive to Jerusalem in his car
Explanation:	There is a match between the prediction and the observation.

Hence, the structure of the mentalistic explanation is like the structure of the Hempelian explanation that I described above. And if this argument is correct, the suggestion that in psychology the explanation assumes the form of "one theory – many models" is mistaken. My counter-argument is that the explanation model argument is incorrect for a variety of reasons, which I shall discuss later on. *One of my counter-arguments is that the above generalization*: if X wants G and believes that B will realize G, then X will do B, *is not an empirical law, but a model, a scheme of a mentalistic explanation*. And as a scheme of explanation, and not as a law or empirical theory, it cannot be set in the premises of the Hempelian scheme of explanation, in the D-N model.

On the assumption that the multi-explanation theory indeed upholds the property of "one theory – many models of explanation", the following question arises: how may one explain behavior by means of this theory, and how may it be tested empirically? To answer this question it is worth examining first how the natural sciences explain complex phenomena, and how one tests theory/ies dealing with complex phenomena (I rely mainly on Bechtel & Richardson, 1993; Simon, 1969, 1973; Wimsatt, 1972). I stress especially the term "complex phenomena" because I am concerned to examine what the implications of the natural sciences are for psychology, which of course tries to explain complex behaviors such as the behavior of Max the cat.

I distinguish simple phenomena, such as the free fall of bodies, which may be explained by an appeal to one natural law, from complex phenomena, such as the eruption of a volcano, how a car, an airplane, or even and electric kettle works, which may be explained by an appeal to several natural laws or theories. For simplicity, we shall look at the familiar working of an electric kettle. How may we explain the boiling of the water? To do so, we must disassemble the working of the kettle into three main subsystems connected with electricity, heat, and water. By means of the laws of electricity we explain how electricity that passes through an electrical conductor, with resistance, creates heat, and by means of the appropriate laws of thermodynamics and chemistry we explain how the heat that is emitted by the resistance rises from the bottom of the kettle and boils the water. Hence, the kettle's action is explained by an explanation of the action of each of its components and by the proper combination of the action of these components, a combination that in the end leads to the behavior that we want to explain – the boiling of the water. This explanation, evidently, is based on a chain of different processes (acting one after the other and simultaneously), all of which are explained by the same kind of mechanistic explanatory model that for its purposes uses various natural laws relevant to electricity, heat, and water. In this respect the explanation of the complex phenomenon in the natural sciences also exhibits the property of "one explanatory model – many laws and theories". The entire difference in the explanation of simple and complex phenomena lies in the manner of application of this property. In the case of simple phenomena, we set every time in the same explanatory model, for example, the D-N model, one natural law matching the given simple phenomenon; in the case of complex phenomena, we set in the mechanistic explanatory model different natural laws, theories, processes that match the components of the complex phenomenon.

Now we move on to the second question: in the natural sciences, how does one examine theory/ies concerned with complex phenomena? I believe that here too the answer is similar to the answer I gave in the case of the explanation "one method of testing – many laws and theories". It is possible to examine separately the theory of electricity, of the emission of heat, of the chemistry of water, and the properties of water, and the way in which these components are combined in the electric kettle unit, by means of the same method of empirical examination, the H-D method. For example, we can calculate the amount of heat produced by electrical resistance and how long it will take the water in the kettle to reach boiling point, and we can test these calculations, predictions, by comparing them with empirical observations.

In brief, then, it seems that in the natural sciences the methodological property "one methodological scheme – many theories" holds when dealing with simple phenomena and also when dealing with complex phenomena. In this light, we shall go back to our previous question: is this the situation in psychology? Does the multi-explanation theory explain, and is it tested like a theory in the natural sciences? My argument is this: *while the test of a theory in psychology takes place as in the natural sciences (by the H-D method) the explanation of complex phenomena is not accomplished as in the natural sciences.* But before we consider the differences in this matter between the natural sciences and psychology, it is worth presenting an example of the explanation of a behavioral episode of Max by means of the multi-explanation theory, and two examples showing that indeed in psychology researchers use the multi-explanation theory to understand several behavioral phenomena.

6.2 Examples from Max's behavior and from psychology

An example from Max's everyday behavior – **Max waits for Aviva**: I have noticed on many occasions that about a minute or more before Aviva enters our apartment, Max pricks up his ears, turns them towards the front door, turns his head, and sometimes even gets down from where he is lying (the sofa, the armchair), approaches the door and sits in his sphinx position before it – waiting for Aviva.

How can we explain this episode? The answer is constructed on the division of this behavioral episode into a number of behavioral components that may be explained as follows:

1) *Max senses that Aviva is on the other side of the door.* It is reasonable to assume that Max heard or smelled Aviva approaching the apartment. These senses in the cat, as is well known, are incomparably better than ours. The explanation of this behavioral component appeals to neurophysiological sensory processes, so in principle this is a mechanistic explanation. However, this explanation, I believe, cannot offer a satisfactory account of the cat's conscious sensory experience (and see chapter 8 on this).
2) *A connection between these stimuli (hearing and smell) and Aviva's mental representation.* The connection between the perception of these stimuli and Aviva's appearance at the entrance of the apartment created a memory, a representation of Aviva in the cat's brain, a representation that has awakened with the receipt of these stimuli. This connection between a sensory reception and Aviva's representation is the fruit of an associative learning process that has recurred time and again over the years. The explanation of this process is mechanistic, and is based on the analogy with the computer as a system that acquires, codes, represents, and operates these representations.
3) *A connection between Aviva's representation and the expectation that she will enter the apartment.* Aviva's representation has aroused in Max, again through a long learning process, the expectation that Aviva is about to come in through the front door; this expectation is accompanied by the highly positive emotions connected with Aviva's image – associations which likewise have been learned over years, coded, and stored in the cat's memory.
4) *Connection of expectation and emotions to approaching the door.* The positive expectation and emotions connected to Aviva's arrival have aroused in Max responses of approaching the door. This response of approaching is at root instinctive, that is, there is a tendency to draw close to positive stimuli (water, food, warmth, i.e., stimuli that arouse a good feeling) and to draw back from negative stimuli (stimuli that arouse pain, fear, i.e., a bad feeling). In the present case Max used the approach response in order to achieve a new goal: eliciting responses of affection from Aviva. The explanation of this behavioral component must therefore refer to these mental components and be integrated into the framework of mentalistic explanations.

Clearly, the present explanation is far more complex than the three-stage interpretation of **scratching armchair-knees**, which was based essentially on a new use of the scratching response. In the present case the episode of **Max waiting for Aviva** is explained by reference to a large number of pieces of information, learned over years, stored, retrieved from memory, and merged with the series of responses (pricking up the ears, turning the head, getting off the armchair, approaching the front door) to achieve a relatively new goal – an affectionate response from Aviva.

May not a simple mechanistic learning explanation be suggested for this episode, instead of the complicated explanation of the multi-explanation theory? For example, the cat has learned to approach the door when the appropriate stimuli appear because this behavior has won positive reinforcement (attention, affection) from Aviva. I do not believe that this explanation is adequate. Being a mechanistic explanation, it is very hard to explain why Max waited for Aviva at the door in only a small percentage of cases. Furthermore, when he pricked up his ears and turned them towards the door, not always did he also get up from where he was lying and go toward the door. If the explanation is mechanistic, we would expect that in all cases (in which stimuli heralding Aviva's arrival appear) the response of approaching the door will appear, because according to the present explanation Max's behavior is analogous to the behavior of a machine. As Keijzer (2001) writes, concerning the purpose of folk psychology as a mechanistic process,

> The goal is to formulate an explanation which does not involve any thinking or sentient agent in its premises. The explanans should involve *no one* who is acting as an intelligent, sentient force, guiding behavior in the right direction. (p. 26)

In other words, if the explanation is mechanistic, we would expect that every time Max receives the appropriate stimuli he will prick up his ears, get up from where he is lying, and walk toward the door. However, Max behaves as Max wishes, and in many cases, even though he has perceived the stimuli, he does not move from where he lies, and sometimes even though his ears have cocked in the direction of the door he has decided, for some reason (like ourselves, I would say) that he doesn't feel like stirring from his place. That is, the present situation matches the condition of free will (see chapter three): the same individual (Max) whose private behavior has changed, in the same state of stimulus (Aviva is about to enter the apartment) responds with different responses (pricks up/does not prick up his ears; approaches/does not approach the front door) at different times.

Two examples from psychology: (a) *Sensory perception*. Assume that Ruth has to take a hearing test. Earphones are placed over her ears and the operator begins to transmit whistle-like signals over a range from sub-threshold level (below the hearing threshold) to the above-threshold level. The purpose of the test is to check what Ruth's hearing threshold is, when this threshold is defined as the strength of the whistle below which Ruth does not hear anything and above which Ruth hears the whistle. Ruth's task is extremely simple: to say if she hears or does not hear the whistle in the earphones. What could be easier than that? If she hears, that is, if her sensory system responds to the strength of the stimulus sounded in the earphones, Ruth will respond positively, and if her nervous system does not receive the physical stimulus of the sound waves, Ruth will respond negatively. The explanation of the phenomenon, therefore, is mechanistic and is based on the neurophysiology of the auditory system. Pure and simple. But is it?

Let us assume that Ruth has been told about the structure of the test: it comprises 200 trials, in some of which a whistle sounds and in some of which a whistle does not sound. (If the whistle sounded in all the trials, all she would have to do is respond positively every time.) Moreover, Ruth also knows that the whistle sounds in seventy-five percent of the trials, namely 150 times. Now, in light of this information, how will Ruth respond? Can we still say that the response (hears–does not hear) will be determined exclusively by the neurophysiological system? I believe not. Ruth certainly will be influenced by the information she possesses and will tend to respond positively even if the whistle is not sounded, but the background sound of the system (a kind of long ssssss) will be heard in the earphones; this is called 'white noise'. She is liable simply to interpret this noise as a whistle.

The important point, which I wish to emphasize in this matter, is that in addition to the physical stimuli Ruth's response is influenced by a large collection of other factors, such as the knowledge she possesses, the degree of her wish to pass the test, her mood, and her attitude to tests of this kind. That is, it will be hard to explain Ruth's test results only on the basis of her sensory hearing system. Account must be taken also of her mental system, her motivation, her emotional state, the significance of the test for Ruth (perhaps the whistle awakens in her memory unpleasant events that happened in her life); in short, account has to be taken of the complex of factors that are not considered pure neurophysiological factors. If this is so, the explanation of "simple" behavior like this must be based on many explanations, on mechanistic as well as mentalistic explanation models. The interesting question, of course, is how may we know how to break Ruth's response down into its different components, into those requiring an explanation by means of a mechanistic model and those requiring a mentalistic model.

To answer this question psychologists developed a special theory called Signal Detection Theory (see summary and discussion in Macmillan & Creelman, 1990). This theory distinguishes two kinds of stimuli: (a) background noise of the system + a signal (the telephone receiver makes background noises above which we hear the speech signals); (b) background noise only. This theory seeks to clean off the background noise and to intensify the ability to detect the signals. In the human system, for example, in Ruth's hearing test, the theory wants to distinguish between Ruth's ability to detect the whistle itself and the different influences (knowledge, motivation, emotion, etc.) on her response – hearing or not hearing. The theory therefore differentiates between two basic processes: the process itself of detecting the signals, which is not influenced by factors of knowledge and motivation (a process based on the sensory system alone), and a process that inclines Ruth to say that she hears more than she does not hear (or the reverse). (This inclination is called 'response bias'.) By virtue of the distribution of answers (e.g., of Ruth) according to four possibilities – presence of the whistle, and Ruth has answered: hears or doesn't hear; and non-presence of the whistle and Ruth has answered: hears or doesn't hear – signal detection theory developed two main indices. One index estimates the degree of detection of the signals themselves, and the other estimates how much the respondent's (Ruth's) response was influenced by factors of knowledge and motivation.

Signal detection theory proposed good explanations for a large number of research fields in psychology, such as sensing (the respondent sensed/did not sense a touch on the back of here hand), discrimination (the respondent discriminated/did not discriminate a visual pattern on a ground), and memory (the respondent identified/did not identify the face she saw relative to other, distracting faces). The important point that I want to stress with this illustration is just one: signal detection theory, to my mind, is an instance of a multi-explanation theory based on the fact that a response to a stimulus is influenced by two kinds of factors. One kind is associated with the nervous system whose action is explained mechanistically, and the other kind is associated with the mental system whose action is explained by means of mental factors.

(b) *Human memory*. The dualist theory of memory is based on the distinction between two kinds of memory stores: short-term memory (STM) and long-term memory (LTM) (e.g., Baddeley, 1976). STM deals with cases in which the individual is exposed to information for a short time (a few seconds), the exposure is once only, the amount of information stored in this store is small (about seven digits plus/minus two), and recall takes place in a short time (up to about 30 seconds).

For example, David wants to phone Ruth and he asks Yossi for her phone number. Immediately after he hears the number David dials it without any mistake. The theory posits that the information is located in David's consciousness, he is aware of it, and is able to recall it at once, as one who still hears the number in his head. But if after hearing the number Yossi goes on to ask David, "So why are you calling her?" David will forget the number and will ask Yossi to repeat it. The reason, as noted, is that the amount of information stored in STM is limited and new information (Yossi's question) that enters STM dislodges and replaces the old information. Humans are incapable of containing in their range of awareness a large amount of information; they can concentrate only on a small and limited number of items.

By contrast, the name of a schoolfellow is stored in our LTM for years without our being aware of it, that is, without this name recurring and entering the STM. This name is not present in us consciously all the time. If this and other details of information were present in our conscious all the time, our awareness would be forever awash with bits of information – a condition that would not contribute to sanity, to put it mildly. Recalling this name requires the action of a special mechanism of information retrieval, an unconscious mechanism, which is not present in the conscious. Think, for example, of the many cases in which you try in full consciousness to recollect the name of a beautiful and admired movie star, and although you know that this name is known to you, you can't pull it out of your memory. This is an embarrassing, disturbing, and irritating situation, and despite all efforts your memory fails you. And then, after you have despaired of remembering the name, a little while later (sometimes a few moments or hours, and sometimes days), this name suddenly floats into your consciousness, makes its appearance in your awareness out of nowhere, enters straight into the STM without your making any effort at all to recollect it, and you know at once that this is the name you so wanted to recall earlier.

As in the foregoing case, the dualist theory of memory is presented here as an example of the multi-explanation theory, which treats a large number of memory phenomena, such as forgetfulness over time, the phenomenon of the serial position, learning pairs, amnesia, memory chunking, and information coding (see Baddley, 1976).

I maintain that these examples illustrate well that complex behavior is based on two kinds of action: action controlled by the will, beliefs and rules of the game known to us; and automatic action controlled by involuntary mechanistic mechanisms. In fact, I would say that almost all the behaviors that can be subjected to ordinary, everyday observation, or a considerable part of the phenomena investigated in psychology, are an elaborate mixture of automatic, mechanistic components, and of conscious, voluntary, and teleological components. Furthermore, many lengthy debates in psychology as a scientific discipline are closely bound up with questions such as whether learning and memory in humans necessitate awareness, consciousness, or whether these phenomena can be explained by means of automatic, mechanistic, non-conscious mechanisms (see, e.g., Rakover 1993). If indeed this is the case (and in my view it is the case), we have no alternative but to try to understand complex behaviors with the aid of the multi-explanation theory.

6.3 Three methodological problems connected to the multi-explanation theory

Although the discussion so far has focused on two kinds, two categories of explanation, the mechanistic and the mentalistic, it is worth emphasizing that each category, each kind, contains several explanation models. For example, the first kind, the mechanistic, contains such models and approaches as the D-N model (Hempel, 1965), the Statistical Relevance (SR) model (Salmon, 1971), the Causal Mechanical model (Salmon, 1984), the Unification model (Kitcher, 1989), and the Manipulation Causation-Explanation approach (Woodward, 2003); and the second kind, the mentalistic, contains, among others, the teleological model and the model of rule-following. In other words, I distinguish two broad categories of models – the mechanistic and the mentalistic, where the mechanistic category comprises a number of mechanistic explanation models associated with various research areas within the natural sciences, and the mentalistic category has a number of mentalistic explanation models associated with various phenomena connected with mental processes and states (e.g., feelings, emotions, thoughts, desires, beliefs, intentions, knowledge, and consciousness). To the best of my knowledge no argument or theory has as yet been found demonstrating that mentalistic models can indeed be reduced to mechanistic models, and that consciousness can be grasped in material terms. The issue takes the form of a debate: for example, while Dennett (1969) supports the approach that it is not possible to perceive intentionality in physical terms, Allen (1992) takes issue with this approach and suggests that intentionality can be understood in terms of a causal explanation. (And see in chapters 8 and 9 below wide-ranging treatment of this subject in support of the approach that mentalistic explanations are different from mechanistic explanations.) Hence it may be proposed that the multi-explanation theory is based on a number of explanation models that do not accord with each other, and as a result several problems arise, for most which I shall offer solutions.

6.3.1 The ad hoc explanation problem

As the multi-explanation theory comprises a large number of explanation models, the possibility exists that this theory can supply in a trivial manner an explanation for every phenomenon on earth. If a certain phenomenon is not explicable by means of explanation model (a), we shall turn to explanation model (b), and if this model is no good either, let's go on to the next model…until the model is found that can explain the particular phenomenon. That is, given a behavioral phenomenon, the multi-explanation theory will always succeed in finding a suitable explanation. On the face of it, this theory is the one to which we aspire, the ideal theory that explains all. So what is wrong with it? The answer is that this theory is not subject to empirical testing, that is, to empirical refutation. For example, if the theory succeeds in explaining phenomenon A by means of explanation model (a), but does not succeed in explaining a certain experimental change in phenomenon A (phenomenon A'), we shall not infer that the theory is refuted, but we shall seek among the models of this theory another explanation model, for example, explanation model (b), which will be able to provide the explanation for phenomenon A', and so on. That is, the theory is impervious to refutation because the multiplicity of models ensures that no phenomenon will be found for which an explanation will not be found.

6.3.2 The inconsistency problem

As the multi-explanation theory includes a large number of explanation models, the possibility exists that this theory will provide an explanation for a certain phenomenon with the help of explanation model (a), and simultaneously the theory will provide for the same phenomenon an opposite explanation by means of explanation model (b). That is, the theory will predict that a certain behavior will take place by means of explanation model (a), and at the same time the theory will predict that this behavior will not take place (or an entirely different behavior will take place) by means of explanation model (b). This theory is therefore liable to suffer from internal contradictions: to predict a thing and its opposite.

6.3.3 The incomparability problem

As the multi-explanation theory contains a large number of explanation models, the possibility exists that it will not be possible empirically to compare two multi-explanation theories if they use different models. To clarify this matter we may look at the following case. We assume that two theories, which use different explanation models, offer an explanation for the same phenomenon. One explanation is good and the other is not. That is, one theory predicts the investigated behavior and the other theory is not able to predict that same behavior. Can it be said that the second theory has been refuted? The answer is no, because the second theory may be good but the explanation model that this theory used simply does not suit the phenomenon under discussion. If the two theories used the same explanation models we could say that the second theory was refuted, while the first theory received empirical support. This situation is similar to an attempt to solve mathematically a problem with two unknown variables by means of one equation. In the given case, we need to suggest answers to two questions or to decide by means of one experimental result between two possibilities: A. Are the theories effective? B. Do the explanation models suit the phenomenon under consideration? (For a discussion of the latter issue see next chapter.) That is, it cannot be determined by means of one experimental result what to blame: the theory (the first or the second) or the explanation model (used by the first or the second theory). (The present problem may be seen as part of the problem known as Duhem's problem, whereby it cannot be known what to blame – the theory or auxiliary hypotheses – in a case where a prediction derived from a theory and auxiliary hypotheses is refuted by observations. See Duhem, 1996 and a proposal for a practical solution in Rakover, 2003.)

The distinction between the theory and the explanation model used by the theory is not simple, and some of the best minds have failed in the attempt. Rakover (1990) showed that the debate between Hull's learning theory and Tolman's theory, which proceeded on the level of running experiments (e.g., learning a T maze), where Tolman wanted to show that his experimental results refuted Hull's theory, is in fact a hidden theoretical debate over which explanation model is better suited to deal with animals' (white rats') learning in the laboratory: a mechanistic model or a cognitive-teleological explanation model. While Hull tried to explain learning by a rat as a machine, as the behavior of a robot-rat, Tolman tried to explain learning by means of cognitive representations, a cognitive map of a maze, and the rat's anticipation of obtaining reinforcements at a certain place in the maze (e.g., on the right side of the T-maze). These two renowned researchers did not pay attention to the fact that methodologically observations support/refute hypotheses, but not explanation models.

As I mentioned at the beginning of the chapter, in addition to these three problems the multi-explanation theory entails two more: how this theory proposes explanations, and how it can be tested in an empirical experiment. However, the discussion of these two problems and the attempt to solve them will be illustrated better after I suggest the solution to the three present problems (see discussion in Rakover, 1997). In the setting of this solution, which forms the conceptual infrastructure of the multi-explanation theory, I will be able to present more lucidly the solution to the two additional problems. The reason is this: if it is not possible to solve the three problems, it is not possible to propose an explanation by means of the multi-explanation theory, nor is it possible to test it empirically.

6.4 Guidelines for the solution of the three problems

To solve these three problems, the following steps should be taken:

A. ***Matching***: The match between the explanation model and the investigated behavior must be determined in advance. As every behavioral phenomenon may be broken down into behavioral components, for each of these the explanation model that suits it must determined. I call the matching pair of behavior and explanation model the "explanatory unit or module";

B. ***Organization***: In the multi-explanation theory all the explanatory models have to be arranged coherently. That is, the multi-explanation theory must be composed harmoniously of several explanatory modules.

After I describe these two procedures I shall explain how the three problems are solved.

First I shall expand on the problem of **matching** and then I shall move on to discuss the problem of **organization.** Hempel maintained that in science (including the natural and social sciences and the humanities) there is one kind of explanation model, but it turned out that this opinion is incorrect (see discussion in Dray, 1966; Salmon, 1989; Woodward, 2003). His model does not suit the social sciences and the humanities; furthermore, even in the natural sciences it is hard to apply his model to a field such as biology. As a result, and on account of the fierce criticism of his approach, which I shall not dwell on here, researchers and philosophers proposed alternative explanation models (see Salmon, 1989; Woodward, 2003). That is, even in the natural sciences several models of a mechanistic explanation prevail, so naturally the question arises as to how in the natural sciences the models and the investigated phenomena may be matched. The match in the natural sciences is fairly easy because an entire domain of phenomena is matched to one explanatory model. For example, Newtonian physics is explained by application of the Hempelian model (the D-N model), while neurophysiological phenomena require an explanation model that details the causal mechanism that produces the neurophysiological response under study (see Hempel, 1965; Salmon, 1989; Schaffner, 1993).

And what happens in psychology? Sadly, here it is not possible to match an explanation model to an entire domain of behavior. For example, it is not possible to suggest that perception is explained by means of one explanation model and social behavior by means of another explanation model, because almost every behavior is composed of mechanistic and mentalistic elements, and the explanation therefore requires a multi-explanation theory. In psychology one must examine each and every behavioral phenomenon, break it down into its components, and match the relevant explanation models to them. As this examination depends on a large number of factors (theoretical and empirical knowledge), I am unable to propose here a formula for the solution of the question of matching, but only guidelines.

6.4.1 How to determine a match between an explanation model and a given behavioral phenomenon

As I argued above, we feel that a phenomenon requires an explanation when a change occurs in a given behavioral situation. When we discern a change in behavior, we at once ask what has caused this change. As long as the behavior continues as before, in its habitual way, no cognitive need arises to explain this behavioral situation. Newton stated that a body will continue in its straight movement as long as a force does not act on it. But if a change takes place in the speed of the body's motion the need for an explanation immediately arises. According to Newton, a change in motion means that a force is acting on the body. That is, we explain the change in the behavior of the body by means of a certain cause – force (e.g., gravitational force). So also in psychology. A change in behavior calls for an explanation. The question is of course which explanation: physical? Physiological? Cognitive? Mental? Emotional? Social? How may we decide which explanation model to choose? I believe that the answer may be given by means of a three-stage process, of which the first stage is the most important:

a. It has to be determined if the given phenomenon requires the use of a mechanistic or mentalistic explanation model.
b. If we conclude that the explanation model is mechanistic, we must clarify what is the most suitable specific mechanistic model.
c. If we conclude that the explanation model is mentalistic, we must clarify what is the most suitable specific mentalistic model.

The *first guideline for matching*, then, is to check if the given phenomenon (i.e., a change in behavior) belongs to the domain of the mechanistic explanation or the domain of the mentalistic explanation. The answer to this question is determined by means of several procedures, and their use offers good enough guidance for achieving the purpose of the matching.

a. It is possible to examine if the given behavior fulfills the conditions developed above for mechanistic or mentalistic behavior: the criterion for mechanistic behavior, the principle of new application, and the criterion for mentalistic behavior.
b. Heyes & Dickinson (1990) proposed a criterion for an intentional explanation whereby an individual's action would not take place in the absence of an appropriate desire or belief. For example, in their view, an intentional explanation cannot be attributed to a given behavior if it can be shown experimentally that this behavior does not change according to a change in the conditions of the environment, a change that is expected to change the individual's belief. As an example, these authors describe an experiment showing that chicks continue to approach a food dish even though their approach causes the dish to move away from them, while moving away from the food dish causes the dish to come closer to them (and see discussion and debate on this criterion in Allen & Bekoff, 1995; Heyes & Dickinson, 1995).
c. Pylyshyn (1984) suggested examining empirically whether a behavioral phenomenon exhibits cognitive impenetrability. If a behavioral phenomenon does not exhibit cognitive penetrability, namely it is not affected by a change in the goals of the individual, in her beliefs, her desires, her thoughts, and her knowledge, it is reasonable to suppose that this phenomenon is based on automatic processes tailored in the brain from birth. As an example, we shall look once more at Müller-Lyer's illusion:

   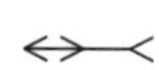

   Although we know that the right-hand line is equal in length to the left-hand line (take a ruler and measure), our perception says clearly that the right-hand line is longer than the left-hand one. That is, this perceptional phenomenon is not influenced by our knowledge. Furthermore, this illusion is found in fish and in chicks too (e.g., Coren & Girus, 1978). So it is reasonable to assume that the explanation of this illusion requires a mechanistic explanation model associated with the brain's neurophysiological structure.
d. The indices offered by Signal Detection Theory (which I described above) may be used to test whether a certain phenomenon is influenced by factors that are not physical or physiological, that is, by factors of motivation and knowledge.
e. As I argued earlier, in humans the question of the link between the given behavior and mental factors can be broken down into the following two questions (and see discussion in Rakover, 1993):
   a. Is a human likely to be aware of her investigated behavior?
   b. Is this awareness likely to influence this behavior?

   In my opinion, these two questions are partially independent of each other. On the one hand, we are not aware of an enormous number of processes that take place in our body and our brain. Not only are we not aware of various chemical and electrical processes, we are not even aware at a given moment of all the information (indeed, an infinite amount) that we have learned in our lives. Relative to what is imbedded in our brain, I will not be mistaken in stating that we are aware of a rather small and limited amount of information. This information too can be divided into different levels of awareness. For example, as you read this passage, most of your attention is dedicated to the text, and very little attention is directed to the surrounding stimuli. Therefore, these phenomena, located basically outside our consciousness, call for an explanation by means of a mechanistic explanation model (whether physiological or cognitive).

On the other hand, even if we are aware of a certain phenomenon, this knowledge does not always exercise a direct effect on the given behavioral phenomenon. Therefore, in these cases too, as I illustrated with the Müller-Lyer illusion, it will not be efficient to use a mentalistic explanation model. The explanation, in fact, will have to tackle the question of how the illusion is created by mechanistic means (computational and neurophysiological) and also the question of perceptual awareness, which apparently arises at the end of the non-conscious processes.

f. Neurophysiological research may offer an answer to the question whether one may explain a given phenomenon mechanistically or mentalistically. For example, Johnson (1972) shows that several aggressive responses are controlled directly by an electrical stimulation in the cat's brain (in the hypothalamus) and also in the brain of other animals (e.g., monkeys); and LeDoux (1996) shows that an important part of fear behavior (immobility, hair bristling, change in blood pressure) appears without awareness and is controlled directly by a brain area called the amygdale. These findings call for mechanistic explanations.

The *second guideline for matching* is linked to the second stage in the three-stage process for choosing the explanation model, the choice of the suitable mechanistic explanation. As the professional literature covers these models in minute detail, I shall not expand on them here (see extensive discussion in Hon & Rakover, 2001; Hempel, 1965; Rakover, 1990; Salmon, 1989; Schaffner, 1993). All I shall say is that the researcher has to decide if the investigated phenomenon is amenable to the best and most efficient explanation according to the following (theoretical and empirical) considerations:

1) Whether the explanation is connected to the occurrence of the behavior itself or to its probability. In addition to the D-N model, Hempel (1965) proposed two other models to handle probability phenomena: the Deductive-Statistical (D-S) model, which is very similar to the D-N model but includes in its assumptions a statistical law and it predicts deductively the probability of the given phenomenon; and the Inductive-Statistical (I-S) model, which is like the D-S model but includes in its assumptions a statistical generalization and it predicts inductively the probability of the phenomenon. (The approach of these three models is called

'the covering-law theory', because in the three models – D-N, D-S, I-S – well-founded empirical laws or generalizations are used to cover the phenomenon under investigation explanatorily.) Salmon (1971) suggested an alternative explanation model to the I-S model which does not rest on a statistical law but on relevant statistical relations. For example, we suggest that factor B is relevant statistically to the explanation of phenomenon A, where the conditional probability of A given B [p(A/B)] is greater than the probability of A [p(A)]. (In fact, Salmon required inequality of the probabilities, but for the purpose of the present discussion it seems to me that the relation p(A/B) > p(A) is intuitively perceived better than inequality. See discussion on these matters in Salmon, 1989; Psillos, 2002; Woodward, 2002, 2003.)

2) Whether the explanation is connected to the kind of explanation models that answer the question **why**, that I call *a fit to a theoretical framework* (or *fitting-scheme*), or to a kind of models that answer the question **how**, that I call *a mechanism (process) of production* (or *production mechanism*. See discussion on this matter in chapter 9). Are we concerned to show that the behavior is a particular case of a general law, as in Hempel's models (i.e., fitting-scheme), or to show how the behavior is created as a result of the action of a certain mechanism or process (i.e., a mechanism of production)? For example, do we want to show that certain learning by Danny or by rat no. 17 is a particular case of a general law such as the law of effect (whereby the probability of the occurrence of a reinforced response will increase)? Or do we want to show that this learning, and its like, are created as a result of the action of a certain neurophysiological or cognitive process, mechanism, which contains concepts, representations, and processes that act on these representations?

The *third guideline for matching* for the choice of the specific mentalistic explanation is of course connected to the behavioral phenomena that we are convinced are influenced by mentalistic factors. Which mentalistic explanation model shall we choose? For a possible answer, we shall study two behaviors clearly influenced by certain information and by a motivational state:

1) Dan stopped his car at a red light.
2) Dan stood on a chair and took down from the top shelf Kant's *Critique of Pure Reason*.

How may we explain these two behavioral phenomena? (And see discussion on this matter in the previous chapter.) We shall look first at the first behavior. Clearly, Dan's driving the car is explained by means of the traffic regulations practiced in Israel, that is, acquired information. Drivers must stop at a red light. And Dan, as a good driver, follows these rules and carries them out properly. These rules are not a kind of natural laws, for the simple reason that they can be infringed with the greatest of ease. If Dan doesn't "feel like" obeying these rules, he will run a red light. And if he is caught by the police he will be punished. By contrast, it is impossible to suggest that a stone falling from the

Tower of Pisa will decide out of the blue to remain suspended in mid-air. Nor is it at all reasonable to suppose that Galileo would punish such a stone for its improper behavior – behavior that does not accord with his law. A large part of human behavior can be similarly explained – behavior explained by a model of rule following, whether these are legal rules, social or religious norms, rules of behavior of a certain group of people, or private rules of behavior that people establish for themselves.

Now we shall look at the second behavior. It is hard to explain this behavior by means of rules because there is no rule that states that a book is taken down by one's standing on a chair. Standing on a chair is a response made to realize Dan's purpose – to acquire wisdom by reading Kant's book. But standing on the chair is nothing but one of a large number of responses that may well bring Dan's wish to fruition. He may, for example, go up a ladder, or ask a tall person to hand him down the book. This behavior, as I noted earlier, is therefore explained by use of an explanation model called the "teleological or purposive explanation": mounting a chair is a kind of behavior likely to realize the purpose of the performer of the behavior.

Although these two models of explanation have a wide area of behavioral overlap (see previous chapter), the models differ in several respects. First, as stated above, not every behavior has an accepted rule. For example, learning a new behavior is not based on existing rules, and often we need the teleological model to explain how new rules of behavior are acquired.

Second, although most rules of behavior have accepted purposes (see traffic regulations) in many cases we uphold rules of behavior as befits decent people, without understanding the reason for this behavior. Why is it obligatory to behave in such and such a way, the child asks; because that's the right way, replies the father. You'll understand when you get older. And why in this religious ceremony is it necessary to behave in such a way? Because that's how a pious person acts, that's what is written in the book of commandments.

Thirdly, like computer hardware that follows the rules of computing determined in the program, without the hardware having the slightest idea of what the purpose of the computation is, so the brain carries out certain rules without the possessor of the brain understanding the purpose of these calculations, without our being aware of it. The aim, if it exists at all, is explained by means of scientific theories, such as the theory of evolution.

These differences, to my mind, point to a certain independence of the rule and the purpose: there are rules that have clear and defined purposes, rules intended to realize specific goals; there are rules whose purpose is very vague; and there are rules without purpose, and they are followed automatically.

My third guideline in the present context is therefore connected to an attempt to determine whether for a certain behavior it is more efficient and suitable to explain it by means of rules with clear and known purposes or by means of rules whose purpose is vague or non-existent.

The *fourth guideline for matching* – the *principle of explanations matching*: The match between an explanation and a behavior must take into account the research fact that a behavior can be divided into several behavioral components. This fact raises the following question: What is the relation between a type of explanation that has been matched to a certain behavior (A) and types of explanation matched to the components of this behavior (a_1 a_2 a_3 etc.)? To solve this problem I propose the "principle of explanations matching":

(a) Behavioral components of a mentalistic behavior, that is, a behavior that is explained by a mentalistic explanation model, are likely in part to take various mentalistic explanations and in part various mechanistic explanations;

(b) Behavioral components of mechanistic behavior, that is, behavior that is explained by a mechanistic explanation model, will only take various mechanistic explanations.

In other words, I propose that components of mechanistic behavior cannot take mentalistic explanations, while components of mentalistic behavior are likely to take mentalistic or mechanistic explanations. I believe that this order of matching constitutes a methodological principle which conforms with the three foregoing guidelines and with the discussion conducted in the previous chapters. For example, if the Müller-Lyer illusion is not influenced by mentalistic factors such as knowledge of the structure of the illusion, the phenomenon requires the use of a mechanistic explanation model and it is hard to see how different cognitive elements of this phenomenon are likely to take a mentalistic explanation. And the reverse: realization of a goal, for example, a journey to Tel Aviv from Haifa to meet Ruth, can be broken down into several components some of which require a mentalistic explanation (e.g., traveling according to the traffic regulations) and some of which require a mechanistic explanation (e.g., seeing Ruth's face). Hence, if mechanistic behavior A is broken down into two behavioral components, a_1 a_2, where a mechanistic explanation is matched to a_1 but a mentalistic explanation is matched to a_2, then either behavior A was not purely mechanistic or the match of the explanation model to the behavioral component a_2 was not successful. (Here is the place to note that matching the kind of explanation to the behavioral component does not solve the problem of interaction between mentalistic behavior and mechanistic behavior. This is a kind of mystery, which the multi-explanation theory tries to contend with partly by use of the organization guidelines which determine the order of activity of explanatory units: see below.)

As is evident from this discussion, the match between a behavioral component and the explanation model is a question in which theoretical and empirical considerations abide together, and it has no straightforward solution. It is worth highlighting this matter, and saying that there is no cause for alarm from this conclusion. The matching process indeed weighs heavily on the work of the researcher in the social sciences, but it is inevitable. One must become accustomed to it, as we have become accustomed to operational definitions. The operational definition is a necessity in every piece of research, because it determines the suitable connection between the theoretical concept and the observation. This connection, as every first-year psychology student knows, is based on relevant theoretical and empirical considerations. For example, the operational definition of aggression is connected to considerations about what is regarded as an expression of aggression in theory, in the professional literature, and in that culture wherein the research is being conducted. What I suggest here, in fact, is that thought has to be given not only to the question of the operational connection, but also to an additional kind of connection, which also is based on theoretical and empirical considerations, the match between the kind of explanatory model and the behavior.

6.4.2 How should the explanatory units be organized?

The *first organizational guideline* and the most important in the organization of the explanatory units is this: avoid as much as possible breaking the 'explanatory unit', that is, splitting the behavioral component-explanation model pair. After the researcher has succeeded in building this explanatory unit with recourse to theoretical and empirical considerations, she must continue with her research on this basis. This guideline does not say that it is utterly prohibited to split this unit. All it suggests is that to break the unit the researcher has to find weighty reasons to suggest a new, better and more efficient explanatory unit in its place. If the researcher inclines to break up the explanatory units with great ease, she will not only not solve the three problems above, she will aggravate them.

The *second organizational guideline* is that the explanatory units must be organized in a theoretical framework in a coherent way that will not lead to self-contradiction and to violation of the above 'principle of explanations matching'. I think that the following two organizational structures will not bring about loss of coherence: a. the chain structure; b. the ramified structure. (And see a discussion on the breakdown of a given phenomenon into its components and their organization in different structures in Bechtel & Richardson, 1993; Simon, 1969, 1973; Wimsatt, 1972.)

a. *The chain structure*: The multi-explanation theory may contain a number of explanatory units and can suggest a certain order of activation among them. This theory is likely to treat different aspects of a behavior and to assume, for example, that explanatory unit (2) goes into action after explanatory unit (1) goes into action. For example, the dualist theory of memory assumes that information first enters the STS and then passes to the LTS. As I noted earlier, since the information in the STS is present in consciousness, it is reasonable to suppose that the explanation model appropriate for explaining the action of this memory store will not be based on a mechanistic model. For example, if David wants, he can rehearse the material in his consciousness endlessly and as a result remember this information until his final hour. And as the information in the LTS is not present in awareness, it is reasonable to suppose that the appropriate explanation model will be mechanistic. As may be seen, this organization, which suggests an order of action of the explanatory units, by its very nature, prevents complications and self-contradictions, which may arise from non-organized and uncoordinated activations of the explanatory units.

The theory may also suggest an explanation, a process that deals with the way in which information passes from one memory store to another. For example, the dualist theory proposes that information present in the STS passes to the LTS when new information enters the STS and displaces the previous, old, information, from the STS directly into the LTS. The question is how shall we characterize this process of passage of information? As the process is unknown in our consciousness, and takes place automatically in our everyday life, I suggest that its explanation may be accomplished through a mechanistic model. As the dualist theory developed entirely by analogy with the computer, it is worth characterizing the passage of information by means of terms and processes taken from the world of the computer. In this case, evidently, the memory explanation comprises three explanatory units: mentalistic (STS), mechanistic (passage), and again mechanistic (LTS). (It should be emphasized that these units too are amenable to further breakdown, for example, in accordance with processes of coding and retrieval.)

Finally, it should be noted that a theory of this kind may greatly expand, and include chains made up of a large collection of explanatory units; moreover, often we must base the structure of the theory on several chains (of different size) acting in parallel, where some of these chains have points of contact (through different explanatory units) among themselves.

b. *Ramified structure*: The multi-explanation theory can offer, to begin with, an explanation for a behavioral episode by use of one explanatory model. However, as we are concerned to deepen and expand the explanation (i.e., to answer additional questions), there will be no choice but to break down the given behavior into different components of behavior, which will require matching of different explanation models. As an illustration we shall look at David's behavior, as he waves to Ruth, who is going to fly off in an airplane, as a gesture of parting. As David is likely to vary the parting gesture according to the information he possesses, for example, he may shout "Au revoir" or blow a kiss into the air, we shall seek to explain the behavior by means of a mentalistic explanation: a teleological model. David waves because he believes that this is the behavior that will realize his purpose – to part from Ruth. Is this an adequate explanation? If *this is all* we seek from the explanation, the answer is affirmative. But if we want to enlarge and deepen the explanation, we shall press on and ask: why precisely the wave? On what grounds does David think that Ruth is indeed Ruth? By what does David know that Ruth has grasped his message? The answers to these questions require us to break down David's behavior into several components, that is, to several behavioral ramifications.

Let us examine the following three possibilities: David needs (a) to identify Ruth; (b) to be sure that waving is in fact a gesture acceptable to and understood by

Ruth; (c) to know that Ruth has indeed received his message. Each of these ramifications requires a match of an explanation model. (And each of these behaviors may be broken down further, a course that will come to a halt the moment our explanatory-epistemological curiosity is satisfied.) Without entering into a detailed analysis of every behavioral ramification, it is clear enough that these three behaviors require different models of explanation. The first behavior, identification of a face, takes place in everyday life automatically without our paying any attention to it. (Here I disregard the fact that apart from the visual pattern of the given face, we are assisted by many additional clues to make the identification, for example, information about the person's gait, dress, age, sex, etc. See discussion in Rakover & Cahlon, 2001.) Treatment of this kind of behavior (based on the visual stimulus alone) therefore needs a mechanistic explanation model. The second behavior, waving as a gesture of leave-taking, requires the use of acquired social information (David must assume that this elementary signal is indeed known to Ruth as a member of the culture to which he belongs), and so also the third behavior: David must receive feedback from Ruth that indeed she has received the message, for example, she waves back. These two behaviors, as they depend on learning of social norms, require treatment with mentalistic explanatory models (e.g., rule following).

6.4.3 Do the guidelines help solve the three problems: an ad hoc explanation, inconsistency, and incomparability of theories?

The multi-explanation theory, based on two kinds of explanation models (mechanistic and mentalistic), may provide us with an explanation for the investigated behavior better than a theory based on one kind of explanatory model, simply because this explanation will be more comprehensive and will deal with the two behavioral aspects together: the physical-biological aspect and the experiential-mental-social aspect. Still, the explanation of this theory cannot reach perfection because an explanatory lacuna will remain that is very hard to fill: our inability to explain in the framework of the concepts of the natural sciences the mysterious mind/body connection (and see discussion on this later, especially chapter 8). And in our recognizing this advantage of the multi-explanation theory we must now probe whether this advantage is not eliminated in the attempt to solve the three problems.

I argue that the obligation of **matching** the behavior component and the explanation model, and the obligation of **organization** of the explanatory units, solve these three problems, because for each behavioral component the researcher uses one single explanation model and one single order of structure of the multi-explanation theory. As a result the researcher cannot propose any explanations for the behavioral component, nor can she leap from model to model as she likes, but she must use one model determined as the most suitable to treat the kind of phenomena under investigation. This obligation prevents the possibility of suggesting ad hoc explanations. The explanation model is determined in advance, as in the methodology adopted in the natural sciences. There, for example, it is clear that one must address the motion of bodies by means of an explanation model of the kind proposed by Hempel (1965).

Matching and organization also protect the researcher from the charge of incoherence. As every phenomenon has a matching explanation model determined in advance, and as the order of the parts of the explanation of the given phenomenon is set in advance, that is, the order of joining the relevant explanation units in the framework of the multi-explanation theory is fixed in advance, a situation will not arise in which the researcher will use different models for the same behavior, nor will a situation arise in which the researcher will decide on a different explanatory order as the fancy takes her. As a result, a situation will not arise of self-contradictions that is liable to arise from the use of different explanation models or from a change in the order of activation of the explanatory units.

Matching and organization allow the researcher also to perform an empirical and theoretical comparison among different theories. This comparison could be complete if two multi-explanation theories have the same structure of explanatory units. However, even if the structure does not fully match, the researcher may still partially compare the explanatory units common to the two theories.

On the assumption that the guidelines of matching and organization indeed help in preventing these three problems, it is now worth going back to the question of whether sound construction of the multi-explanation theory also helps resolve the two theoretical questions we alluded to above: explanation and empirical test.

6.5 Multi-explanation theory, giving an explanation, and empirical test

Earlier I suggested that while a theory in psychology is tested essentially as in the natural sciences, that is, by use of the same H-D method, the explanation of complex phenomena is not done as in the natural sciences. Now, in light of the description of the multi-explanation theory (which solves the three above problems), I shall discuss this proposal in greater detail.

6.5.1 Is the multi-explanation theory tested by use of the H-D method?

The answer is affirmative. The testing method is indifferent to the kind of model used by the researcher to explain the results of the experiment. As long as a certain prediction can be derived from the theory under consideration in a given experimental situation, and as long as the prediction can be compared with the observation, the H-D method can be operated. Furthermore, the method of the empirical test is amenable to application to the theory as a whole or to parts of it, that is, to the explanatory unit that the researcher wants to test empirically. (Here I bypass the so-called Duhem's problem which states that an isolated hypothesis cannot be refuted. See discussion and practical proposal for a solution to this problem in Rakover 2003.)

We shall look at David's behavior of waving as a gesture of parting from Ruth. The explanation of this behavior is constructed on the teleological model: David believes that waving will be perceived as a sign of parting. This explanation can be tested in several ways. For example, we can ask David if he knows Ruth. If the answer is no (Who on earth is Ruth?), we can reject the explanation that the waving is a gesture of leave-taking from Ruth. We can press the point: is he parting from someone else? No, David replies. I haven't said goodbye to anyone. Then why did you wave? Oh, says David, recently I've been having nasty twinges in my right shoulder and I found that swinging my arm eases the pain.

Should David answer that he does know Ruth, we can test part of the multi-explanation theory. If David knows Ruth, he can surely identify her picture too. To examine this hypothesis empirically, we lay out before him a series of pictures of ten women, and ask him to choose Ruth's. If, for example, David tells us, after peering at the ten pictures, that Ruth's picture is not among them, we have refuted the hypothesis that David identified Ruth and waved to her in parting. Is Ruth's picture really not here? we ask David in amazement. No, David replies. Then who did you wave to? Oh, says David, to my little niece Ruthy who boarded the plane with her parents.

In short, essentially I see no obstacle to suggesting that a multi-explanation theory is amenable to an empirical test as performed by scientists in the natural sciences. The question is, why do I use the word "essentially"? The answer is linked to the differences we found between the methodology practiced in the natural sciences and that practiced in psychology (see chapter 4): first, in a large number of theories in psychology the connection between the terms of the theory, between the assumptions and the conclusions, is not constructed on a logical deductive or mathematical basis, but on the basis of pragmatic inferences made in everyday language; and secondly, the concept-observation connection in psychology is fairly loose.

6.5.2 Is the explanation offered by the multi-explanation theory similar to the explanation offered in the natural sciences?

The answer, to my mind, is negative. First, while complex phenomena in the natural sciences take explanations resting on mechanistic schemes of explanation, the explanations proposed by the multi-explanation theory in psychology rest on two schemes of explanation: mechanistic and mentalistic. I call this the *mentalistic difference*. Secondly, while in the natural sciences the explanations for complex phenomena (including dynamic phenomena) are mechanistic, which treat a given system as if functioning like a machine in the same way over time, in psychology account has to be taken of the change, the mental development of the individual. I shall call this the *mental development difference*.

The *mentalistic difference*:

Bechtel & Richardson (1993) propose explaining complex systems by means of 'mechanistic explanations', and this is how they define these explanations:

> By calling the explanations *mechanistic,* we are highlighting the fact that they treat the systems as producing a certain behavior in a manner analogous to that of machines developed through human technology. A machine is a composite of interrelated parts, each performing its own functions, that are combined in such a way that each contributes to producing a behavior of the system. A mechanistic explanation identifies these parts and their organization, showing how the behavior of the machine is a consequence of the parts and their organization. (p. 17)

(See a similar approach to mechanistic explanations in Bunge, 1997; Carver, 2001; Machamer, Darden & Carver, 2000.) Here it is worth noting that I use the term mechanistic explanation in a broad sense, to include the present explanation and also other explanations based on natural laws (according to the Hempelian approach), causal models (e.g., according to Salmon's (1984) approach), genetic models, mathematical models, models of the computer, and so on. (On 'functional analysis' see also the previous chapter and Cummins, 1983, 2000.)

As an example of a complex system (but simple in comparison with a complex system such as the weather), we shall revisit the explanation of the action of the flashlight. How does this instrument work? To answer the question we perform the following theoretical-empirical analysis.

1) *Decomposition into parts*: We decompose the instrument – the flashlight – into the following parts: switch, battery, electric wire, bulb.
2) *Explanation of every part*: The explanation of every part is based principally on the laws of electricity. When the switch is operated, the electric circuit is closed. The current passes through the filament in the bulb (e.g., a tungsten wire), heats it, and as a result the bulb emits light. And now we approach the explanation of the action of each and every part: the action of the battery on the basis of chemical and physical processes, the action of the filament as electrical resistance, and finally laws of electricity concerned with connection in series as distinct from connection in parallel. All these explanations are based on known physical theories (electricity and chemistry) and are controlled by mechanistic explanation models.
3) *Explanation of the interaction between the parts*: We explain the connection between the different parts as a process of conversion of energy: chemical energy which converts into electrical energy which converts into heat which emits light. This energy conversion can be calculated precisely and explained by means of the appropriate laws, theories, which are set in the appropriate mechanistic models.

In what does this explanation differ from the scheme of the explanation that I call the multi-explanation theory? To consider the difference I shall propose here three properties important for the explanation of the flashlight.

a) Explanatory independence: We break down the overall system into its components, its parts, and explain each part separately. For example, how the battery or the bulb works in a flashlight, or how the engine, the electrical system, or the cooling system works in a car. We assume that each and every part is independent in the explanatory aspect. That is, we assume that despite the important interaction between the parts, an interaction that determines the action of the system as a whole, each and every part, each and every component, can be given an appropriate and adequate explanation.

b) Explanatory relevance: For example, regarding the action of a car, each and every part in the overall system produces a number of products; for example, the motor causes the mechanical motion that rotates the car's wheels, but it also generates noise and noxious gases. For the explanation of how the car works as a single entire unit, the noise and the gases not only are irrelevant, they are a kind of negative products whose adverse effect must be reduced to a minimum. What is important for the action of the car is the relevant fact that the engine converts energy associated with burning benzene into mechanical energy that can be harnessed for the purpose of setting the car in motion. That is, although every part is amenable to an explanation independently of another part that contributes to the overall system, each part has an important property that is relevant to the proper action of the system as a whole.

c) A uniform kind of explanations: The explanation that we described above is based on diverse physical theories that are applicable in the specific case of the flashlight too. The unifying property of these theories is that they are all mechanistic explanations. This property allows us to attain a uniform explanation of the entire system. This goal is realized by the rules of transformation, by means of which we pass from one computation of energy to the next. That is, these rules of transformation allow us to perform a precise calculation of the results that are measurable empirically, so that we can know exactly the amount of current that will pass through the filament in the bulb and as a result we know the amount of heat and light created in the bulb. That is, the use of these theories allows us to calculate precisely the amount of energy of each and every component, and the order of the conversions of energy produced by the battery to the energy produced by the flashlight bulb.

Considering the foregoing, the argument that I propose is that these three properties are not realized fully in the multi-explanation theory. The reason lies in the use of mechanistic and mentalistic explanation models together. If every behavior were amenable to explanation by means of mechanistic models alone, there would be no essential difference between the methodology of the explanation of the natural sciences and of the social sciences (e.g., psychology). However, as I have stressed again and again, it is not possible to explain the behavior of an animal (including the human) only by means of mechanistic models. The mechanistic explanation for a complex system in the natural sciences is founded on causal processes (conversion of energy) that underpin the interactions between the components of the system (e.g., Bechtel & Richardson, 1993; Carver, 2001; Machamer, Darden & Carver, 2000). It is not so in psychology.

I don't think that today there is any philosophical approach or scientific theory that is universally accepted and that does not spark sharp criticism that is able, in the framework of the scientific game-rules, to bridge the mental and the physiological-behavioral, even though this connection of mind/body is a common and characteristic phenomenon in animals. (And see chapter 8 on this matter.)

In mechanistic models the phenomenon being investigated is predicted with logical, mathematical, probabilistic, or causal-physical necessity. By contrast, in mentalistic explanation models, the phenomenon under study is predicted by means of practical (private) considerations, made by the behaving animal or human.

For example, David can behave according to the traffic regulations, and then we shall say that David is behaving rationally. But let us change the situation a little: David is driving fast at three in the morning down empty streets because he is in a great hurry to get to the hospital where, on account of a sudden and massive heart attack, his mother has been admitted. Now, shall we not justify David's speeding above the allowed limit? Is David behaving rationally? On the one hand, he is breaking the law, endangering his life, and thus reducing his chances of seeing his mother; on the other hand, the chances of an accident at this early hour are very small, while the chances of his reaching the hospital after his mother has died are, sadly, high. The decision to drive fast or slow is therefore subject to David's considerations, beliefs, and feelings. That is, what is relevant depends on David's viewpoint.

The behavior of animals and humans, as stated, is complex and requires a mechanistic and a mentalistic explanation. As a result it is hard to materialize the property of explanatory independence as it is realized in the natural sciences because the same motor movement may assume different meanings. Furthermore, it is not clear how the conversions of energy between diverse behavioral components may be calculated. That is, it is not clear how to calculate the interaction between the elaborate explanatory unit based on the mechanistic explanation and the mentalistic explanatory model.

In this respect, compared with the explanation given in the natural sciences, the explanation offered by the multi-explanation theory is partial and not complete. In the case of the flashlight the explanation constitutes a single whole (because the explanation details the conversions of the energy, that is, the mechanism that provides the light when the switch is activated and closes the electrical circuit). By contrast, in the case of David's waving as a gesture of taking-leave of Ruth, we are unable to detail wholly the mechanism responsible for his behavior from the moment David catches sight of Ruth's face until the moment he performs the given behavior.

Despite these differences, it is worth noting that in the natural sciences too there are complex phenomena that are difficult to comprehend by means of application of the 'decomposability' strategy: breaking down the parts of the system, and identifying the functioning and the organization of the components. In fact, phenomena may be aligned on a scale from simple phenomena (such as the flashlight) to the most complex ones (such as the human brain). Phenomena of the first kind are given to decomposability and easy identification of the functioning of their components, for example, in the case of the flashlight. But phenomena of the latter kind are typified by a great multiplicity of interactions among the components, a situation liable greatly to hinder the application of the decomposability strategy and understanding of their action. Yet not all researchers think this way. For example, Barendregt (2003) discusses an approach whereby it is not possible to break down very complex behavioral phenomena into their components, and therefore it is not possible to offer a genetic explanation for these behavioral components. He takes issue with this approach and recommends a way that leads to giving a mechanistic explanation even to such complex behaviors. In this respect, it may be said that in comparison with research in the natural sciences the work of the scientist attempting to understand the behavior of animals, involving the mind/body interaction, is infinitely harder.

The mental development difference:

Bechtel & Richardson (1993) suggest two complementary strategies for an explanatory decomposition of a complex system: the analytic strategy, whereby a physical component in a system is isolated and its functioning is examined, and the synthetic strategy, whereby the researcher proposes a scheme of behavior consisting of different parts as a hypothesis for the functioning of the system, a hypothesis that is tested empirically. While the analytic strategy stems from the bottom-up approach (i.e., from an attempt to understand a system on an explanation track that goes from the basic components to the structure of the whole system), the synthetic strategy stems from the top-down approach (i.e., from an attempt to understand a system on an explanation track that goes from the structure of the whole system to its basic components). As can be seen, these two strategies helped us to understand the action of the flashlight. Can these two strategies be applied to the behavior of Max the cat? To answer this question, we shall take another look at the behavioral episode of **scratching armchair-knees**, which was explained by means of the three-stage interpretation as follows (see chapter 5):

(A) The general framework of the explanation – the purposive explanation: *Max* wishes to get *petting from me* and he believes (by virtue of previous learning, acquisition of knowledge) that *scratching* the armchair will win him *petting*, therefore *Max scratches* the armchair.

(B) Explanation of the scratching response: This response in itself is explained by an appeal to a mechanistic explanation, that is, to the anatomical-physiological structure of the claws and their evolutionary function: hunting, defense-attack, replacing the claws and marking the furniture (in the Rakover apartment) with the cat's smell.

(C) Integration of the scratching response into the purposive explanation: How is the scratching response, whose explanation according to (b) is mechanistic, integrated into the framework of the purposive explanation? The integration is brought about by a learning mechanism, that is, Max has

> learnt that under certain conditions the scratching response leads to realization of his purpose – petting from me. Now, in many cases where the desire arose in Max to be petted by me, he used this acquired knowledge to attain his goal. This goal (petting) is entirely different from the previous functions achieved by the use of the claws: while the previous functions were connected to hunting, defense-attack, replacement of claws, and emitting the cat's smell, the new purpose is different – petting (stroking, tickling, and other expressions of affection).

As may be seen, in this explanation I have used both strategies, the analytical (isolation of the scratching response) and the synthetic (organization of the entire behavior by the teleological explanation), together. But here an important fundamental difference enters between the decomposability explanation proposed by Bechtel & Richardson and the three-stage interpretation. While the decomposability explanation concerns a complex phenomenon that functions in the same way at different times, like a machine (e.g., a flashlight, a car, an airplane, a computer), the three-stage interpretation treats Max's behavior by focusing on the mental development and change that took place in his behavior, namely the change to Max's new goal, which he achieved by means of the same response, the scratching response. In this respect the three-stage interpretation is a dynamic explanation, which describes a mental development expressed in the behavior episode of **scratching armchair-knees** by focusing on the function change that occurred in the response component of the scratching: from a survival-evolutionary function to a new function of achieving petting. This change goes right to the heart of the teleological explanation, in that Max of his own free will scratched the armchair because he believed that scratching the armchair would realize his new purpose – petting.

By contrast, analysis of the functions of the components in a mechanical system remains fixed: the batteries will always produce electricity, the bulb will always produce heat and light. Moreover, even if we test more perfected machines, learning machines able to store information and alter their behavior in different situations, in the end it will become clear that these are nothing but machines that function according to the laws of chemistry and physics. For example, even if we improve our flashlight so that it will change the power of the light according to the state of illumination round about, we will get nothing other than a slightly more sophisticated mechanical system than a regular flashlight, which contains some kind of light sensor that changes the power of the flashlight's illumination, a sensor that always acts in the same way. (Of course, still more perfected machines may be suggested, constructed, for example, on the basis of neural networks, machines likely to amaze us even more with what they can do; and perhaps, ultimately, we shall build 'Robocat', which will imitate all Max's deeds, a machine that will astonish us endlessly, but will remain, after all, only a machine. And see the discussion on this matter of building a human robot devoid of consciousness, a 'zombie', in chapter 8.)

CHAPTER 7

Establishing multi-explanation theory (a)

The mentalistic explanation scheme

The goal of the present chapter is to establish methodologically the multi-explanation theory according to the scientification approach: to show that specific teleological explanations (will/belief) are produced from a mentalistic teleological explanation scheme, which functions according to the game-rules accepted in science. Just as specific explanations in the natural sciences are based on mechanistic explanation schemes (e.g., the D-N model), so specific explanations for the behavior of humans and animals are based on a mentalistic teleological explanation scheme. To justify this argument I show that this explanation scheme maintains several methodological properties of explanation schemes in the sciences, such as the property that the refutation of a specific mentalistic explanation carries no implications for the explanation scheme itself. Furthermore, I show that the teleological (will/belief) explanation scheme is not a kind of scientific law, because it does not maintain characteristics acceptable for a law in the sciences. The chapter also compares the teleological explanation scheme with other explanatory approaches: Dennett's "intentional stance" and Cummins' "functional analysis", takes issue with them, and considers the differences between them and the present approach.

A noble person is one graced with sublime qualities A, B, and C, but I found them also in a sewer rat.

The basic argument that I put forward in the last chapter is this: while in the natural sciences the same model uses laws and various theories to explain diverse phenomena, in psychology the same theory uses several explanation models – mechanistic and mentalistic. I called this the multi-explanation theory. In light of this distinction, the following question arises: is the multi-explanation theory a kind of scientific theory? In my view the answer is yes. In this chapter and the next I shall try to provide grounding and justification for this answer. In the present chapter I shall show, according to the scientification approach, which seeks scientific legitimacy of explanations assumed in everyday life (see an extensive discussion on this approach in chapter 9), that specific mentalistic explanations are produced by mentalistic explanatory schemes, which in large part meet scientific methodological requirements. Science, as grasped in the natural and social sciences, explains different phenomena by the use of explanation schemes that uphold certain methodological properties. Here I shall show that mentalistic explanation schemes too uphold a considerable share of these methodological properties, therefore it will be possible to add these schemes to the inventory of procedures contained in scientific methodology, even though these schemes cannot be perceived as mechanistic.

In the next chapter I shall discuss the argument that to date no accepted philosophical approach or scientific theory has been found for reducing the mind to the body, or consciousness to the brain. This discussion is important because if consciousness can be conceived in terms of natural science, the methodological basis of the multi-explanatory theory is undermined.

7.1 A model, a mentalistic explanation scheme

To answer the question, does the source of a specific mentalistic explanation lie in a mentalistic explanation scheme that is scientifically acceptable, we have to clarify what are the characteristics of scientific explanatory schemes, and to check the extent to which these characteristics also apply to mentalistic explanation schemes. To the extent that these schemes do not maintain these characteristics, it will be hard to treat them as part of science. As we shall see later, the mentalistic explanation scheme does maintain these characteristics.

In a large proportion of cases we understand human behavior or the behavior of animals by an appeal to a mentalistic explanatory scheme, which uses internal mental factors. In general, we use the following four mental schemes:

1) *Will (motivation) and belief*, where we want to provide an explanation for purposive behavior (i.e., we use the teleological explanation scheme);
2) *Rules of behavior*, where we want to provide an explanation for normative behavior (i.e., we use the rule-following scheme);
3) *Cognitive ability (thinking, logic)*, where we want to provide an explanation for the use of information and rational processes of inference;
4) *Emotional state*, where we want to provide an explanation for extreme, and usually not rational, behavior.

We ascribe these (and similar) psychological factors-reasons to the individual and through them explain her public behavior. For example, why did Ronny pull the armchair under the light in the lounge? Because he wanted to change the bulb that had burnt out (mentalistic scheme (1)); why did Ronny get up and offer his chair to the elderly man? Because Ronny acted according to the norm of "show respect for your elders" (mentalistic scheme (2)); why did the policeman lie in wait for the escaping crook precisely at the railway station between Haifa and Tel Aviv? Because he inferred from the crook's behavior that he would leave the train at the midway station and not at Tel Aviv station (mentalistic scheme (3)); and why does David live only in ground-floor apartments? Because he suffers from fear of height (mentalistic scheme (4)).

In this chapter I shall concentrate on the teleological model alone (scheme (1)) for the following reasons (see chapter 5). Although for every mentalistic model it is possible to draw its characteristic field of application, it is fairly easy to show that the teleological explanation is likely to apply also to fields of behavior addressed by other mentalistic models. For example, David wishes to travel from Haifa to Tel Aviv to meet his girlfriend Ruthy, and believes that a journey in his car *obeying the rules of the road applied in the state of Israel* will bring his aim to fruition. Therefore, David drives according to the accepted regulations. As behavior according to the traffic regulations is part of the rule-following explanation model (mentalistic scheme (2)), it transpires that the present teleological explanation (scheme (1)) also refers to behavior according to rules.

In light of the literature discussing the question of explanation, I suggest that the explanation model, the scheme of giving an explanation, has five major features (see discussions in Hempel, 1965; Lipton, 1992, 2001a; Nagel, 1961; Psillos, 2002; Rakover, 1990, 1997; Salmon, 1989; van Fraassen, 1980; Woodward, 2002).

a) General procedures: An explanation model is a scheme or general procedure by means of which the researcher offers a specific explanation, an understanding of the investigated phenomenon. The specific explanation itself, therefore, is perceived as a particular case, as one instance that realizes the general scheme. This property is maintained in the mentalistic explanation too: the specific explanation that David waves in order to part from Ruth is a specific instance of a teleological explanation scheme according to which an individual will perform a certain act if he believes that this act will realize his desire.

b) Factors and reasons: As I mentioned earlier, the cognitive need for an explanation arises when a change in behavior takes place. The explanation model in the sciences assumes that one of the components of the explanation for the change will be linked to a general law, a theory, a process, a certain mechanism on account of which, by reason of which, this change has occurred. That is, the change, the particular event, is an outcome of the general law. For example, the force of gravity explains the change in speed and direction of the artillery shell, and electro-chemical factors explain the flexion of muscles. However, when human or animal behavior is at issue, as stated above, the explanation is accomplished by an appeal to internal mental factors that explain-reason the investigated behavior.

c) Expectation and prediction: The explanation model furnishes us with a certain basis (logical, mathematical, rational, causal, analogical, justificatory, explicatory) that in certain conditions the investigated phenomenon is expected to occur or that the probability of its occurring is likely to increase. For example, we expect that the apple will fall on Newton's head when the justification for this expectation is based on the universal law of gravity. Five hours after breakfast there is a very good chance that David will say that he is hungry and will look for a restaurant. Also regarding the teleological explanation: we expect that given David's wish to take leave of Ruth and his belief that waving is the proper response to realize his wish, it will be only logical for David, as a rational person, to wave. Justification for this expectation, that the waving response will take place, is not based on logic, on statistical probability, or on necessity stemming from a causal natural law, but on reasoning, on likelihood underpinned by social cultural meaning and by the individual's practical evaluation (and see on practical inferences Graham, 2002; Mele, 2003). In all these cases there is justification that the given behavior indeed will occur, due to the force of gravity, hunger, and the wish to part, in the knowledge and belief that waving the hand will realize this wish.

d) Empiricism: The specific explanation that realizes the explanation model must be connecting to reality and allow an empirical test of the theory, the law, the mechanism placed in the explanation model, for example, the D-N model that I described in the last chapter. This requirement applies in the context of the mentalistic explanation too. For example, if we explain the behavior of waving as a sign of a wish to part company, we can test this explanation by presenting certain questions to the waver: Did you wave as a gesture of leave-taking or to ease the pain in your shoulder? Do you know the person you waved to?

e) Indifference rule: As I showed above (chapter 6), explanation models (schemes) and the method of empirical testing are not affected by the results of the experiments, by observations. Empirical results affect only the laws or theories, which are used by the explanation models and the testing method. I propose that a similar rule of indifference applies to the mentalistic explanations: the empirical results exert no effect on the mentalistic models and testing method, but only on the specific explanations, the specific hypotheses, generated by the mentalistic explanation model when it is applied to the behavioral phenomenon under study. To understand this proposal, consider the following example. David wishes to meet Ruth in Tel Aviv and believes that his wish can be realized by taking a bus. Hence, we may propose the specific hypothesis that given David's wish/belief, he will travel to Tel Aviv. However, David does not travel to Tel Aviv. According to the proposed indifference rule, the specific hypothesis is refuted, but not the testing method and the purposive, will/belief explanation model (if X wants to achieve G and believes that behavior B will realize G, then X will do B) on the basis of which the specific hypothesis was generated. This proposition calls for an extensive discussion. It has two main goals: to show that we keep using the purposive explanation scheme, despite the fact that in several cases specific purposive hypotheses are refuted, and to show that the will/belief explanation scheme does not function like a law or a theory in science.

7.1.1 A teleological explanation model and folk psychology

The discussion of the purposive explanation model is conceptualized as an important part in the framework of the philosophical discussion of folk psychology (see on this subject Churchland, 1981, 1988, 1989; Fodor, 1987; Gordon, Morris,1986; Stich & Nichols, 2003; Stich & Ravenscroft, 1994). Folk psychology refers, among other things, to people's ability to provide in a variety of ways explanations for their behavior and of the behavior of other people by an appeal to mental states and processes. For example, why am I so very happy, because I have won 1000 shekels on the national lottery; why did David travel to Tel Aviv, because he wanted to meet Ruth. Folk psychology, then, does not just offer explanations for a behavior that has already happened, it also suggests predictions for a behavior that is going to happen. For example, what will happen to David when he finds out that he has been fired? He is liable to go into a state of depression.

Folk psychology has a number of interpretations that I cannot discuss here in detail (see discussions in the literature cited above). One of the popular interpretations that is highly relevant to our discussion suggests that folk psychology is analogous to a scientific theory. That is, folk psychology has properties typical of a scientific theory, and like the scientist, who uses a scientific theory to explain and predict, the individual uses this psychology to explain or predict people's (and animals') everyday behavior. This interpretation proposes that people have an everyday theory of behavior, and the name usually given to this interpretation is "theory theory" (TT). So TT posits that folk psychology has a structure analogous to the structure of a scientific theory: it comprises mental states (such as feelings, emotions, thoughts, images) that have the status of theoretical concepts in a scientific theory, are in (causal) connection with each other, with stimuli that act on the senses, and with responses, the individual's behavior. The theoretical (explanatory), empirical, significance of these concepts is determined, therefore, according to their function in the theory.

Another interpretation of folk psychology, which in many respects is similar to TT, suggests that this psychology consists of a collection of folk rules, "folk wisdom", platitudes: for example, people in a painful condition will act to assuage this pain; people behave in a certain way because this behavior fulfills their wishes and their belief. These rules function like scientific laws in the explanation and prediction of behavioral phenomena.

A contrary interpretation to the scientific theory approach expressed in the two foregoing interpretations suggests that folk psychology is not a kind of scientific theory but should be seen as a process of mental simulation. (On mental simulation theory (MST) see Gordon, 1986; Stich & Nichols, 2003; Stich & Ravenscroft, 1994.) This approach proposes, for example, that when David tries to explain or predict Ruth's behavior in a certain situation he uses the following procedure: first David gets into his head the state of stimuli, the input data, in which Ruth is, and second David allows his cognitive system to run and to process these data and to produce an output, a response. This response is David's prediction and explanation of Ruth's acts.

These interpretations of the nature of folk psychology and its relation to philosophy of consciousness sparked many debates in the relevant literature, which also involved empirical data from various experiments, and which I cannot discuss here (see, e.g., Nichols, 2004; Stich & Nichols, 2003; Stich & Ravenscroft, 1994).

Several philosophers (e.g., Churchland, 1981, 1988, 1989; Stich, 1983) argue that compared with cognitive psychology or neurophysiology, folk psychology is a spurious science at root, and in the end it will disappear from the book of science, along with all its notions, just as folk theories about ghosts disappeared. Here is what Churchland (1988) writes:

> ...folk psychology is not just an incomplete representation of our inner natures; it is outright *mis*representation of our internal states and activities. Consequently, we cannot expect a truly adequate neuro-scientific account of our inner lives to provide theoretical categories that match up nicely with the categories of our common-sense framework. Accordingly, we must expect that the older framework will simply be eliminated, rather than be reduced, by a matured neuroscience. (p. 43)

Churchland's position, as may be expected, stands in opposition to other interesting perceptions (e.g., Fodor, 1987; Horgan & Woodward, 1985; Kitcher, 1984). I do not accept it either, for the following reasons. Unlike the MST approach, which rejects the interpretation of the scientific theory for folk psychology, I suggest that folk psychology may well be part of the scientific approach. Still, the interpretation I offer differs from the scientific theory approach to folk psychology: I do not accept the viewpoint that the concepts of this psychology (will, belief, intention, knowledge, feelings, emotions, etc.) have roles in laws or in scientific theories, and suggest instead that these concepts have roles in the framework of the mentalistic schemes of explanation of everyday behavior. For example, the use of the teleological explanation (by means of the concepts will/belief), which constitutes an important part of the explanation and prediction of behavior in the theory of folk psychology, is not a kind of law or scientific theory, but a scheme for explanation by means of which specific teleological explanations for specific behaviors are produced. (Some researchers, such as Churchland, 1988; Horgan & Woodward, 1985; Rosenberg, 1988, phrased the teleological explanation in the form of a scientific law. Rosenberg, who discusses profoundly the teleological explanation as a law in the social sciences, suggests, "This then is the leading explanatory principle folk psychology offers us" (p. 25).) The question, of course, is what the justification for my claim is. The answer is in two parts: the first considers the principle of refutation and the second considers the issue of the law in science.

7.1.2 Teleological explanation and refutation

According to the refutation criterion (Popper, 1972/1934), scientific hypotheses, laws, and theories must have the quality of refutability, otherwise they become faith. If folk psychology is part of the world of science, its laws must be subject to the criterion of refutation. I argue that these laws, this folk science theory, are not subject to refutation, not because they are bad science but because they are schemes, procedures, for giving explanations. To show this, we shall examine two instances that are explained teleologically.

Shimshon fell ill with influenza. He wanted to be cured, and believed that inhaling fresh air at dawn would cure him immediately. Nevertheless, Shimshon did not get out of bed at sunrise. David wants to see an entertaining movie this evening, and decides to go to the cinema, but instead of going he stays home. In both these cases the prediction about the individual's action is refuted, for example, the prediction from the hypothesis of "going to the cinema": David did not go to the cinema. Furthermore, when David is asked why he didn't go to the cinema that evening he has no adequate explanation, but still he assures us that he will go tomorrow.

My argument is that this specific hypothesis, the "going to the cinema" hypothesis (and to the same extent the "influenza" hypothesis), has been refuted, but the general scheme for this kind of explanations, the teleological explanation scheme (if X wants G and believes that B will realize this aim, then X will perform B), was not refuted, just as the refutation of the "law" of free fall of bodies $D'=1/2GT^3$ carries no implications for the Hempelian explanation model and the testing method. What is refuted is the law itself (in our case T has to be squared, not raised to the third power) and not the explanation scheme and the testing method (the D-N model and the H-D method). And as in this case of the Hempelian explanation model, although the specific teleological hypothesis, the "going to the cinema" hypothesis, has been refuted, we continue to use the teleological explanation scheme to explain dozens of other specific cases, which for the most part are supported (i.e., the individual indeed behaves according to her will and belief). As this argument is important, it is worth going into in detail:

a. The teleological model is applied to the "going to the cinema" behavior;
b. As a result of this application a specific explanation (hypothesis) is created for David's going to the cinema:
   1. David wants to see an entertaining movie,
   2. David believes that going to the cinema will fulfill his wish.
   3. *Prediction*: David will go to the cinema.

   This explanation, as may be seen, is teleological, and done according to the teleological model, namely this is a specific explanation molded according to the teleological model or scheme.
c. An empirical observation attests that David did not go to the cinema. Hence, the specific explanation "going to the cinema" failed, and the specific hypothesis that David in this case will indeed fulfill his desire was refuted. (However, if the observation attests that David did go to the cinema, this hypothesis is confirmed and the behavior 'going to the cinema' is explained by an appeal to David's will/belief.)
d. The proposal is that what has been refuted is the specific hypothesis, and not the teleological explanation scheme and the testing method themselves. The purposive explanation model continues to create specific explanations, hypotheses, which are expressed by further observations of David (and of other people) showing that David (in most instances) does behave according to his will/belief. If this

> were not the case it would not be possible to put any specific purposive hypothesis to an empirical test, because in principle one negative result (the lack of a single match between the predicted and the observed) would be enough to refute the specific hypothesis, the testing method, and the purposive explanation model, which generated this hypothesis.

My argument that it is not possible to refute the scheme for explanations of actions by will/belief, because this scheme is not influenced by contradictory results, is of course amenable to alternative interpretations, which attempt to explain why it is so difficult to refute purposive explanations. These interpretations, as we shall see, do not properly deal with the case under study. Non-refutation may well stem from a number of reasons, such as:

(1) Teleological explanations are not a kind of law or empirical generalization acceptable in the natural sciences because these explanations have flaws and drawbacks, for example, the argument that a conceptual, logical, connection exists among the concepts will, belief, and action;

(2) Laws and theories in science are not amenable to proof and refutation either.

Answer to (1): Rosenberg (1988), who summarizes relevant literature on the subject and deals in great detail with (1), formulates teleological explanations in the form of a law:

> [L] Given any person x, if x wants d and x believes that a is a means to attain d, under the circumstances, then x does a. (Rosenberg, 1988, p. 25)

Later he goes on to discuss, among other things, the argument that the connections between the concepts in [L] are logical:

> [L]'s functions are to show us what counts as having a reason for doing something and to show us when a movement of the body is an action. Thus, desires, beliefs, and actions are logically connected, not contingently connected, by [L] and therefore not causally connected by [L] or any causal law. (p. 37)

Furthermore, the suggestion that behavior is a function of desire and belief in fact constitutes an equation with two unknowns: the behavior shows us what the animal desires on the assumption that we know what it believes; and the reverse: the behavior shows us what the animal believes on the assumption that we know its desire (see discussions on this subject, (1), and on those linked to it, in Allen, 1992; Bennett, 1991; Rosenberg, 1988; Sayre-McCord, 1989).

In contrast to causal laws in the natural sciences, in the teleological explanation it is not possible to measure the concepts desire, belief, and action separately. For example, David's action – taking a book off the shelf – is not merely a motor movement but a movement carrying meaning, an expression of his desire (to read a book) and of his belief (that taking the appropriate book off the shelf will fulfill his desire). By contrast, in Galileo's law of free fall of bodies it is possible to measure separately the length of time the body has fallen and the distance it has traversed in this time. If this characteristic of the teleological explanation, [L], holds, it emerges also that it is not possible to refute a specific teleological explanation (for a critique of the argument of conceptual logical connection see, e.g., Rosenberg, 1988; Sayre-McCord, 1989).

As an example we shall peruse again at the explanation of "going to the cinema": David indeed wanted to go to the cinema, but he stayed home. Has the explanation failed? Is the prediction that David will go to the cinema refuted? Let's take a look. If indeed there is a logical connection between the concepts desire, belief, and action, it transpires that a change in action necessitates a matching change in desire and belief. That is, David stayed home because his desire and belief changed accordingly. If this is in fact the case, then staying home is not in the nature of *contradictory* evidence to his intention to go to the cinema, but simply *supporting* evidence of his intention to stay home. Hence, it is not possible to refute specific teleological hypotheses either, because performance of action (2), which is contrary to desire/belief (1), in fact is explained by an appeal to desire/belief (2), which have replaced desire/belief (1).

Although this claim cannot be entirely disregarded, I don't believe that it holds for the following reasons. First, it is not true that separate observed information, separate indications, on the three concepts of desire, belief, and action cannot be obtained. For example, David says: I want to go to the cinema tomorrow evening. The next morning he looks at the newspaper and decides: This evening I shall go to the Or cinema where a film is being screened that I wish to see. In the evening David gets into his car and drives to the Or cinema. Similarly, when Max wants to get off the knees he begins to look right and left, his body tenses, and then he jumps off the knees. That is, it is possible to obtain information attesting to the individual's intention by way of a different behavior, which does not overlap with the behavior that realizes the individual's desire. Therefore, one behavior may be seen as an expression of a desire, another behavior as an expression of belief, and action as a meaningful-motor behavior, that is, a third behavior. These comments may pave the way also to a practical experimental solution to the problem of the equation with two unknowns by the receipt of extra information about will/belief in different situations. Similarly, Bennett (1991) suggests that belief is to be inferred from behavior (output), and also from the effect of the state of stimulus (input):

> … the anchoring of beliefs not only to behavioral outputs but also to sensory inputs gives us an independent grip, putting gravel under our feet so that we don't skid uncontrollably. (p. 43)

Secondly, in many cases even though the individual continues to cling to a certain desire/belief, he does not perform what seems to us logical, rational. In the given case of "going to the cinema" David has no satisfactory explanation for his not going to the cinema, even though he promised that he would go the following day. (David does not say that a powerful intention formed in his heart to stay home in order, say, to read a chapter of Kant's *Critique of Pure Reason*. "I don't know why, but I just didn't go. Maybe I didn't feel like it". Furthermore, David can argue that he decided of his own free will to stay home. Why? "Just like that. That's what I decided".) As I have already noted above, and as we shall see later, the connection between an action and desire and belief is not binding, essential, but is a practical connection (see Millgram, 1997).

Thirdly, let us assume a "flat" man, with only one wish in his heart, W, and one belief, B, so that it is possible to realize his wish by performing a certain action, A. It is thus reasonable to presume that according to W&B the flat man will perform A. Therefore, should A not come about, it would be possible to see this as a refutation of the *hypothesis* concerning the flat man. The point I want to stress here is this: the procedure of empirical testing examines first if the observation matches the prediction, and only if there is no match, that is, when the refutation of the hypothesis is obtained, does the question of what happened arise. Why? If we were certain in advance that the observation would not match the prediction, we would not trouble to test the hypothesis empirically. Only after the refutation has been obtained does the question arise of what is to blame: the hypothesis itself, auxiliary hypotheses, the course of the experiment, etc. (see Rakover 1990, 2003). This process also occurs in the present case: only if the flat man does not carry out his intention do we raise such questions as: did he weaken in his purpose or was his mind deflected from it? Did he get drunk? Maybe despite all a new wish arose within that overtook him (i.e., the man is not flat after all); perhaps a consideration appeared before his eyes showing him that A was hard to do; maybe he decided thus of his own free will; and so on.

Answer to (2): The argument that an empirical hypothesis cannot be proven is logically well founded (see discussion in Rakover 1990). The argument that an isolated empirical hypothesis cannot be refuted, a condition known as Duhem's problem, states that an empirical test of a given isolated hypothesis is done by means of auxiliary hypotheses, so it is not clear what is to blame (the hypothesis or the auxiliary hypotheses) when the results contradict the prediction derived from the hypothesis together with the auxiliary hypotheses. Duhem's problem set off a great debate in the literature, and I offered a practical solution to it, which accords with the practice of conducting research in psychology that tests isolated hypotheses empirically (Rakover, 2003). In this sense, I do not tend to accept the general argument of non-possibility of refutation of an isolated mechanistic hypothesis as well as the non-possibility of refuting a specific teleological hypothesis.

So I find that both these interpretations of the non-possibility of refuting a teleological hypothesis do not hold. Here I must stress yet again that I do not claim that a *specific teleological hypothesis* cannot be refuted, only that a *teleological explanation scheme or model* by means of which we create specific hypotheses cannot be refuted.

7.1.3 What is a suitable explanation scheme?

In consequence of the present discussion the following question arises: if counter-observations controvert the specific explanation but not the explanation scheme, the explanation model, by what shall we determine that a given explanation scheme is proper and suitable? I discussed a part of the answer at the beginning of this chapter, and the answer to this question is likely to carry us to an interminable discussion of the subject of scientific explanation, which is far beyond the aims of this book; nevertheless, I shall substantiate the intricacy and the complexity of the matter with the help of the following points (see discussions on this issue in Hempel, 1965; Lipton, 1992, 2001a, 2001b; Nagel, 1961; Psillos, 2002; Rakover, 1990, 1997; Salmon, 1989, 1992, 2001a, 2001b; van Fraassen, 1980; Woodward, 2002, 2003).

The discussion in the literature on providing scientific explanations entails broad metaphysical and epistemological approaches, by means of which attempts are made to answer the question of what is required of a scientific explanation. Some of these approaches enjoy agreement and some do not. For example, despite the argument that observations do not refute an explanation scheme but laws, theories, and hypotheses, the requirement arises that a scheme must allow a comparison between the explained and the observed, that is, that the explanation scheme must permit an empirical test (e.g., Hempel, 1965). For example, in the D-N model use is made of a scientific law connected to empirical observations by means of which empirical predictions are derived; in the teleological model the concepts desire, belief, action refer to observable events (e.g., by verbal report and observation of a behavior). It is not possible that the explanation proposed has no connection with the empirical observations, or that the explanation itself in no way matches the phenomenon under investigation.

A further requirement is that the information supplied by the explanation will constitute good grounds for expecting that the investigated phenomenon will in fact take place (or has taken place). Indeed, we expect that given the law of free fall of bodies, the body will fall a certain distance in a given time; and given David's desire/belief, it will be rational to expect that David will perform a certain action to realize his goal.

These two general requirements are agreed, but researchers differ over many others. As an illustration, we may look at the demand that a good explanation must offer causes for the occurrence of the studied phenomenon, and at the fact that there are several explanation schemes that are not reducible to any universal scientific explanation scheme. Lipton (1992, 2001a) discusses five concepts, profound intuitions, that furnish us with understanding of the investigated phenomenon. An explanation offers: good *reasons for* believing that the phenomenon took place; understanding by making the phenomenon *known*; understanding by grasp of the phenomenon as a *uniform* part of the whole; understanding that the phenomenon happened *necessarily*; and an explanation that suggests a *cause* for the phenomenon's occurrence. Lipton examines these five concepts and concludes that causality is the most efficient concept, despite several problems linked to it, for example, philosophers still do not have a good and agreed theory of causality, and in science many explanations are not causal. These drawbacks of the concept of causality notwithstanding, it serves on the one hand as a criterion for good and effective explanation schemes and on the other as an important source for proposing new explanation schemes.

Salmon (1989) in his excellent book summarizes, illumines, and criticizes discussions on the question of explanation, which developed over more than four decades. As an example we shall look at the D-N model. The model has been critiqued in several respects, for example, a legitimate scientific explanation does not have to have the form of an argument; an explanation does not have to be based on the use of a scientific law (e.g., see van Fraassen's (1980) pragmatic explanation model); and the D-N model does not succeed in explaining phenomena that evidently require the use of causality. As a result of these critiques and discussions, alternative models were developed that require that the explanation be based on the concept of causality, for example, the model developed by Salmon (1984), called the Causal Mechanical (CM) model of explanation; and Woodward's (2003) alternative model which was based, among other things, on criticism of Salmon's CM model.

As a final point in this short answer to our question it is worth noting that Salmon (1989), who includes in his book a discussion of teleological and functional explanations, comments that Hempel himself reached the conclusion that it was not possible to reduce the functional explanation (according to which the presence of a certain component in the system is explained by an appeal to this component's ability to function in such a way as to contribute to the survival and realization of its goal) to the explanation schemes that he developed. Salmon sums up the discussion:

> In the correct D-N explanation the explanans is logically sufficient for the explanandum. In the typical functional explanation the explanandum is, given the conditions, sufficient for the explanans. From Hempel's stand-point that is just the wrong way around. (p. 30) [The term explanans signifies the explanatory conditions, and explanandum the phenomenon to be explained.]

7.1.4 A mentalistic explanation model and scientific laws

As I wrote above, a number of researchers have phrased the teleological explanation (i.e., explanation of an action by desire/belief) [L] according to Rosenberg (1988), in a form that matches the formulation of a law in the natural sciences (see also Churchland, 1988; Horgan & Woodward, 1985). The question at the center of this section is: Is it indeed possible to grasp [L] as meeting the requirements of a scientific law? My answer is negative.

First I shall examine whether [L] upholds general accepted and important criteria that distinguish scientific laws from accidental empirical generalizations (see discussions on this subject and other relevant matters in Swartz, 1985; Weinert, 1995; Woodward, 2000, 2003). To substantiate this distinction I shall draw a parallel between a known scientific law, (Newton's) law of gravity, and an accidental empirical generalization, which I shall call "the law of Ruth's parties". This "law" is based on the observation that everyone who was at Ruth's last party had an IQ higher than 130. Then I shall show specifically that [L], as a law in folk psychology, does not uphold several additional properties characteristic of a scientific law (see Rakover, 1990, 1997).

a. *Counterfactual situations*: What will happen if we throw a stone up into the air? What will happen if we discover a new planet? In these cases the scientific law supports a fact that has not occurred: the stone will certainly fall to the ground, and the tenth planet will behave according the Kepler's laws, which may be derived from Newtonian theory. However, it is absolutely clear that the IQ of at least one of the participants at future parties given by Ruth is liable to be below 130. That is, the law of Ruth's parties does not support a fact that has not happened, being counterfactual.

    It is reasonable to suppose that [L] does not support a counterfactual possibility either, because, for example, x is likely to want b more than d, or x is likely not to know how to perform a (the action that will realize his desire) and the many resultant possibilities on account of which x will not do a. (A reminder: "[L] Given any person x, if x wants d and x believes that a is a means to attain d, under the circumstances, then x does a" (Rosenberg, 1988, p. 25).)

    Rosenberg discusses these things and finds that [L] without the addition *ceteris paribus* is false. In other words, to make [L] efficient we must add to it a long list of factors that are held constant, for example, that x will not want b more than d. The problem is that this list of additions is very long, far in excess of what is acceptable in the natural sciences. For example, Galileo's law requires the falling of bodies in a vacuum. (Interestingly, the formulation of the teleological law in Churchland, 1988; Horgan & Woodward, 1985, already included several factors that are held constant.)

b. *Explanatory power*: The scientific law has the power to explain empirical phenomena. Why did the stone fall to the ground? Because of the force of gravity. But on the assumption that the entry threshold to the law faculty is an IQ of 130, would anyone in their right mind admit to this faculty Michael, for example, simply because he attended Ruth's last party? The answer is obvious. Here we should add that the explanatory power of a scientific law stems from its fitting into the broad theoretical-empirical framework. (For example, the law of gravity is connected to Newtonian theory, the laws of Kepler and Galileo to Copernicus, and to an impressive collection of observations and results of experiments.) And how does the law of Ruth's parties so fit? What is its theoretical empirical basis? This law is not grounded in a theoretical-empirical network, like the law of gravity.

    [L] does not have the same necessary explanatory power as the law of gravity: the act moves from theory into practice by means of practical considerations and reasons. Similarly, the theoretical-empirical connections into which [L] fits, it seems to me, are not the kind of firm connections characteristic of the laws of the natural sciences. (See chapter 4 and also discussion of these matters in Churchland, 1981, 1988, 1989; Stich, 1983, and a critique of their approach in Fodor, 1987; Horgan & Woodward, 1985; Kitcher, 1984.)

c. *Universality*: A scientific law must be generalizable beyond time and space. We assume that the law of gravity acted on the solar system and other systems a billion years ago, and will act on these system a billion years hence. Nevertheless, all the laws of science are bounded by certain limits, by a certain physical system, and constitute an expression of abstraction and physical idealization. For example, the Newtonian law of gravity is restricted to terrestrial speeds and to objects that are not of atomic or subatomic size; and the laws of biology are limited to various evolutionary groups, for example, explanations of the amoeba's behavior cannot explain cats' behavior, and certainly not the highly complicated behavior of humans. The law of gravity refers to Earth ideally as a point of mass, without taking into account that Earth is constructed of different layers of mass that are not distributed symmetrically. Yet it is fairly clear that while the law of gravity applies to all the stars in the Milky Way, it would be hard to maintain that the law of Ruth's party applies to the people and the other parties of Ruth and of Ronit and Dorit.

    This conclusion also pertains if we accept Woodward's (2000, 2003) suggestion that in the special sciences (e.g., economics, psychology) there is no point in talking about laws in the usual sense in physics, but about stable empirical generalizations, invariance, generalizations that do not change beyond relevant variables. For example, the matching law, which I discussed in chapter 4, may be taken as a stable empirical generalization that applies to different subjects, different reinforcements, and different tasks (experimental designs). Compared with the matching law, it is impossible to perceive the law of Ruth's parties likewise as a stable empirical generalization because at the next party it will already become apparent that at least one of the partygoers has an IQ below 130 – namely Michael, Ruth's new boyfriend, who on that account was not admitted to the law faculty!

    It is reasonable to suppose, then, that [L] does not meet the requirement of universality or stable empirical generalization, because, as I argued earlier, this law, without the proviso *ceteris paribus*, is false. That is, the law changes extremely over a large number of relevant variables.

So [L] does not satisfy these three criteria, therefore it does not display the properties of a scientific law. Is [L] a kind of accidental empirical generalization? I believe that answer is negative here also, because this law is not like the law of Ruth's parties. For example, while Ruth's law is based on one observation (one party attended, say, by thirty people), the area of application of [L] is vast, and includes observations of different people (and animals), different kinds of desires, beliefs and actions.

Now I shall move on to discuss specifically whether [L], as a law in folk psychology, maintains several more properties typical of scientific laws. (As may be seen, I considered some of these properties above, so I shall assemble them here.)

(1) The concepts desire, belief, action show a certain interdependency, some logical connections. For example, David's behavior is not just a motor movement

but a behavior, an action carrying meaning – David's desires and beliefs. By contrast, in the law of free fall of bodies the distance of the fall is a term measured methodologically independently of the term time in the duration of the fall. In other words, it is hard to reduce teleological terms to causal terms (which require methodological independence of the cause from the thing caused), so it is hard to see the teleological law as expressing causal processes characteristic of most laws in the natural sciences. Nevertheless, as I noted above, information can be obtained on each of the terms of [L] separately despite their great complexity. For example, measuring the wish to go and see a funny movie differs from the measurement of the belief that the trip to the cinema will realize this wish and is different from the very going to the cinema. Still, it is evident that the connection between will, belief, and action is much broader than a logical link, which suggests, for example, that if A is identical to B, the measurement of A is the measurement of B, and the reverse.

(2) The structure of [L] is different from the structure of a law in the natural sciences. I wrote in chapter 4 that the concepts that appear in the laws of science – Galileo's law – represent a uni-dimensional property (time, distance) or a combination of uni-dimensional properties (acceleration, work, energy). By contrast, the three concepts that appear in [L] are names of very complex behavioral categories: there are different kinds and varieties of wishes, beliefs, and actions. Moreover, the explanatory behaviors – will, belief – evince phenomena themselves requiring explanation, and it is hard to see how one may decompose the three behaviors in [L] to uni-dimensional behavioral components. (Furthermore, this multi-dimensional structure creates intricacies that impede an empirical test of the concepts of the law: will, belief, action, and rationality.)

(3) Measurement units for the concepts of [L] – will, belief, and action – do not exist, as they do for laws in the natural sciences. Moreover, [L] does not satisfy the requirement of "equality of units" because as I showed in chapter 4, in the function Action = f(Will, Belief) the combination of measurement units current in psychology for the concepts of will and belief is not identical to the combination of measurement units for the concept of action.

(4) A law in the natural sciences in itself does not constitute an explanation and does not include a procedure of explanation. A law is a functional description among certain variables that satisfy the requirements of mathematics language, the conceptual background in which the law fits, and the theory of measurement. According to Hempel (1965) an explanation is created as a result of following the rules of the game, a scientific procedure, which utilizes the law as one of its important components. The explanation itself is a kind of conclusion from a logical argument that the phenomenon under study is perceived as a particular case of the law. Is [L] also likely to function

in Hempel's scheme of the explanation like any other scientific law? As I argued above, the answer is no. (See the explanation model argument in chapter 6.)

It is not possible to derive logically, mathematically, the conclusion X will perform B from the preambles X wishes to achieve G and believes that B will realize his goal. The reason is that the transition from the assumptions to the conclusion in the case of [L] is not done logically, mathematically, as is the case in the Hempelian explanation, but this transition depends on practical reasoning, on the practical considerations of each individual. Von Wright (1971) suggests that

> Practical reasoning is of great importance to the explanation and understanding of action. It is a tenet of the present work that the practical syllogism provides the sciences of man with something long missing from their methodology: *an explanation model in its own right* which is a definite alternative to the subsumption-theoretic covering law. Broadly speaking, what the subsumption-theoretic model is to causal explanation in the natural sciences, the practical syllogism is to teleological explanation and explanation in history and social sciences. (p. 27)

(In this passage I have emphasized the words '*an explanation model in its own right*' because here von Wright expresses an idea very similar to the fundamental thesis of this chapter: [L] is not a law, an empirical generality, but a procedure, a scheme, a model, for yielding a specific teleological explanation.)

7.2 A scheme of mentalistic explanation and other explanatory approaches

If my arguments hold, the teleological explanation ([L]) should not be seen as a law or theory but as a scheme, a procedure of explanation that suggests specific explanations, specific predictions, for everyday behavior. This explanatory scheme may be regarded as an alternative to several explanation approaches in psychology, so it is worthwhile my comparing it with them. In the present section I shall draw a parallel between the teleological scheme and the two following proposals: the "intentional stance" approach (see Dennett, 1971, 1987) and the "functional analysis" approach (see Cummins, 2000).

7.2.1 Intentional stance

The purposive explanation scheme bears a similar feature to Dennett's (1971, 1987) approach, called the "intentional stance": both approaches suggest interpreting a behavior of humans and animals by an appeal to will/belief. (On the application of the intentional stance to behavior of animals, to cognitive ethology, see Dennett, 1987.) The intentional stance treats a given system (e.g., a person, an animal, an artifact) *as if* it were a rational agent behaving according to its desires/beliefs, and therefore it makes possible prediction and empirical testing of behavior. This attribution of mental states to a given system makes it possible to predict behavior of the system, regardless of whether this attribution indeed holds. What determines things is the effectiveness of the prediction. For example, it makes no difference if a computer is indeed endowed with will/belief, because what counts is the efficiency of the prediction and the ease of use of the intentional stance. Treating the chess-playing computer as a rational being, which wants to win the game and therefore makes the right moves, helps us to predict its behavior well enough, and with greater ease than the attempt to understand its action from the viewpoint of the structure of its software (which Dennett calls the "design stance") or by considering the structure of its hardware (which Dennett calls the "physical stance").

Despite the similarity between Dennett's approach and mine (as stated, both approaches propose using will/belief as explanations of behavior), I maintain that several interesting differences exist. It seems that one of the goals of the intentional stance is to suggest making use of it for easy, effective, and convenient interpretation, and sometimes ready-made interpretation, of behavior, whether this interpretation is correct or not. I can use an unsuitable and incorrect intentional interpretation as long as the prediction of the behavior of the system under study helps me and serves my purpose; I don't need to know how a car is built, but to understand how to operate it, that is, how to use it. In other words, the intentional stance may be seen as a good means of achieving the user's goals – to get by in life. Dennett (1971) writes in the context of the chess-playing computer:

> Lingering doubts about whether a chess-playing computer *really* has beliefs and desires are misplaced; for the definition of Intentional systems I have given does not say that Intentional systems *really* have beliefs and desires, but that one can explain and predict behavior by *ascribing* beliefs and desires to them, and whether one calls what one ascribes to the computer beliefs or beliefs-analogies or information complexes or Intentional whatnots makes no difference to the nature of the calculation one makes on the basis of the ascriptions. (p. 91)

He goes on:

> All that has been claimed is that on occasion a purely physical system can be so complex, and yet so organized, that we find it convenient, explanatory, pragmatically necessary for prediction, to treat it as if it had beliefs and desires and was rational. (pp. 91-92)

Compared with this, the purposive explanation scheme is offered as an explanation scheme that is proper in the social sciences, a procedure of explanation that creates specific teleological explanatory hypotheses, and that leads to an understanding of behavior by an appeal to mental concepts, such as will/belief, which represent real processes in the individual's head. The specific explanation that arises from the mentalistic explanation scheme is no other than a kind of explaining hypothesis which is to be tested against observations and in reference to alternative hypotheses. This scheme cannot be applied to every system or to every behavior, but to those behaviors of animals that fit into private behavior, into mental processes. That is, attribution of will/belief to animals (such as Max the cat) is done according to objective criteria that were developed in earlier chapters: the 'principle of new application', criteria for mechanistic and mentalistic behavior, the 'three-stage interpretation' and the 'principle of explanations matching'. Moreover, as I pointed out in the previous chapter, the mentalistic explanation scheme is part of the multi-explanation theory. This theory is based on the assumption that complex behavior of living beings (humans and animals) cannot be explained by an appeal to one explanation scheme but by use of mechanistic and mentalistic models of explanation. That is, the mentalistic explanation is not a test of a purely practical stance (convenience and effectiveness) but a necessary explanatory component among the other components required for understanding the investigated behavior.

I distinguish the teleological explanation scheme that is not subject to refutation from a specific teleological explanation that is subject to refutation; Dennett (1987) does not make a similar distinction and suggests, following Skinner, that the mentalistic explanation is not refutable, because its logical structure allows a large number of ad hoc interpretations. He proposes, for example, that it is possible to save the specific prediction that Joe will come to class because he wants a high grade, and believes that his presence in class will realize his desire, from the refuting observation that Joe did not show up in class by the assumption that Joe was obliged to do something else urgently that he had forgotten, and so on.

The problem of non-refutability of the teleological (will/belief) explanation scheme is different from Duhem's problem, namely that it is impossible to refute an isolated hypothesis in the natural sciences (physics) because every hypothesis is tested along with a large number of auxiliary hypotheses, so it cannot be known what to blame when the result of the experiment contradicts the prediction: the hypothesis under study or the auxiliary hypotheses (see Rakover, 2003). The main difference is that empirical observations are important in the latter problem but not in the former.

But in the case of a specific will/belief explanation, I believe that another problem arises, beyond Duhem's problem. Clearly, in science it is not possible to refute the hypothesis that under condition A response R will be obtained, if condition A itself is not realized. Similarly, if the will/belief condition (1) of the individual is annulled or replaced by newer and stronger will/belief condition (2), then the prediction from will/belief (1) cannot meet an empirical test, simply because this condition of will/belief (1) is not realized. Therefore, the counter-observation does not contradict the prediction. In physics the experiment or the controlled observation ensures that condition A is indeed realized, but in the case of the teleological prediction there is no guarantee that will/belief condition (1) will indeed materialize: every living thing is a world unto itself. Nevertheless, by my perception this is nothing but a practical problem of inspection and experimental control. That is, I think that conditions of inspection and control can be constructed by means of which it will be possible to assume with a high degree of plausibility that will/belief conditions (1) will indeed be realized at the time of the experiment or the observation (e.g., by verbal report). In cases of this kind I don't seen any reason in principle why it is not possible to refute specific teleological hypotheses.

Dennett's approach sparked many debates. One of the important criticisms stated that the intentional stance is a kind of instrumental strategy, a convenient calculating instrument that produces good predictions on the behavior of a given system, and is not a kind of scientific theory about real processes of will/belief; Dennet (1987, 1995) took issue with this criticism. Although the discussion of this subject is beyond the aims of the book, it is worth stressing yet again that in my approach there is no pure instrumental treatment of the concepts will/belief, but a realistic treatment: the teleological scheme proposes a specific hypothesis about will/belief that reflects processes that have taken place in the individual's brain, and that are responsible for the action she performed. In the present context the difference between the two approaches comes down to the following question: why do you, the scientist, choose to use the particular theory T? The instrumentalist will answer that she chose T because this is the theory whose predictions are the best there are; the realist will answer that this is the theory that best reflects reality. In this regard, it seems to me that while the approach of the intentional stance will tend to accept the first answer, my approach tends to accept both. (Here it is apt to add that Allen & Bekoff (1997) criticized Dennett's assumption of perfect, ideal rationality in the context of his approach to the animal world, and they contrasted it to Millikan's (1984) evolutionary approach.)

7.2.2 Functional analysis and the status of empirical generalizations in psychology

In his article of 2000 Cummins rejects the application of the Hempelian approach (especially the D-N model) to psychology, because, among other reasons, he thinks that laws in psychology are nothing but a description of behavioral phenomena that require an explanation, so they cannot function as an explanatory component in the Hempelian model. In his opinion, the explanation for behavior can and should be given in psychology by a 'functional analysis' of the structure of the behavior; that is, by decomposition of the behavior into its functional components and their arrangement in such a way as to illumine the function and production of the behavior (and see discussion on this subject in the foregoing chapters, particularly chapter 6). Cummins suggests five frameworks, explanatory structures, which allow functional analysis: explanation by means of will/belief, computational symbol processing explanations, explanation by means of neural networks, neurophysiological explanation, and evolutionary explanation.

Since the teleological explanation scheme, which constitutes a part of the multi-explanation theory, rests in many cases on decomposition of the studied behavior into its different parts, it bears similar features to Cummins' approach. From this angle I agree with Cummins, but I differ from him regarding the status of some of the empirical generalizations in psychology: although these generalizations require explanation, I believe that they have explanatory power. I will justify this suggestion by means of the two following arguments: (a) every generalization offers information that supplies an explanation partially, answers to our questions; (b) empirical generalizations in psychology pass from the status of descriptions requiring explanations to the status of explainers by means of broad-based empirical grounding and theoretical anchoring.

7.2.2.1 *Empirical generalization as supplying partial explanatory information*

In this section I wish to show that empirical generalization possesses something more than the fact that it sums up a collection of observations. This something more is an explanatory power; I suggest that empirical generalizations can supply partial explanatory information about the behavior under study. As an illustration, let us look at the following table of observations that was drawn up for a certain behavioral system:

Measurement of independent variable (X)	Measurement of dependent variable (Y)
1	3
2	8
3	15
4	24

These observations were obtained in a controlled experiment replicated a number of times (in the present system and in other systems of the same sort) and each time precisely these data were obtained. Does this table supply us with some kind of explanatory information? I believe it does. Not only is the correlation between these two variables not accidental (the experiment was controlled and satisfied the replicability criterion), it can be known what the response of the system will be under certain conditions X. That is, if we want the explanation of a behavioral phenomenon to satisfy the requirements of non-accident and prediction (in a certain domain), then this table supplies us with this certain explanatory information. Now let's assume in addition that the scientist who investigated this system looked at the data and discovered that a certain order existed there: when X grows, Y grows too! Very interesting. Furthermore, the same scientist also found that these observations could be summed up into a still more precise empirical generalization, as a simple equation: $Y = X(2+X)$. Does this empirical generalization provide us with any additional explanatory information? I believe it does. This generalization sums up the experimental results with great precision; but in addition, with its help it is possible to predict new behaviors, for example, on condition that $X=5$ we will predict that $Y=35$ – a prediction that may be tested empirically. Furthermore, the precise empirical generalization offers us a partial explanation for the following question: how is it that a system of this kind responds to experimental conditions [X=1,2,3,4,5] in such a special way that its response obtains the following values respectively: [Y=3,8,15,24,35]? The answer is this: because there is a specific functional relation between X and Y; because it is possible to map these observations on a certain mathematical system possessing such and such properties; and because this mapping, which connected the observations to the mathematical terms, is accomplished by means of a certain theory of measurement (and see discussion on this subject in chapter 4).

Clearly, more (why and how) questions can be asked associated with this empirical generalization: does this generalization reflect real processes in the system? How is this generalization, this mathematical description, realized in the system under study? What kind of realization is at issue here: mental processes? Cognitive processes? Neurophysiological processes? Chemical processes? Physical processes? The answers are likely also to offer us a way to a solution to the following question: is it possible explain this empirical generalization, that is, to reduce it to a more basic and broader theoretical-empirical system?

As may be seen from this example, the dividing line between a pure description of behavior and a description that supplies explanatory information is not clear-cut, because an empirical generalization is likely to provide different degrees and different kinds of explanatory information linked to what we want to know – to the questions that we ask. In this respect I believe that an empirical generalization supplies us with some minimum of explanatory information – special predictions determined by the generalization form itself. Am I coming out here against the criticism of the theory that prediction is explanation? Let us see.

One of the critiques of the D-N model and the I-S model is that according to these models explanation is prediction (and on further critiques see Salmon, 1989). Several researchers have given examples of how it is possible to predict phenomena without explaining them, and to explain phenomena without predicting their occurrence – examples that run counter to the prediction/explanation symmetry (see Cummins, 2000; Salmon, 1989). For example, it is possible to predict with a barometer that a terrible storm is about to break out even though a barometric change does not constitute an explanation, a cause of the outbreak of the storm; and it is possible to suggest evolutionary explanations without offering predictions of the future, or to explain that X fell ill with a certain disease because he did not get treatment with antibiotics, without being able to predict that X will fall sick with this disease without treatment with antibiotics, because the probability of falling sick with the disease without antibiotic treatment is fairly low (and see discussion on this case in Salmon, 1989, p. 49: (CE-5) Syphilis and paresis.) My answer to these criticisms is as follows.

First, without entering into a detailed analysis of the examples of explanations without prediction, I can say that these and similar examples are not a kind of decisive proof, because without predictions it is not possible to test the explanations empirically, and if it is not possible to test them empirically their connection to science is very loose.

Secondly, the instances of prediction without explanation indicate, in fact, answers to more questions that the generalization does not have the power to supply. For example, while the correlation between barometric level (X) and strength of the storm (Y) offers us answers and explanations to certain important questions, the correlation does not provide an answer to the important question about the explanation of the correlation itself: that's another story.

I think that if we ranked methodologically the explanation as more important than the prediction, a great number of scientific explanations would find themselves greatly embarrassed. For example, Feynman (1985), a Nobel laureate for physics, writes to the readers in his book *QED: The Strange Theory of Light and Matter* the following:

> It is not a question of whether a theory is philosophically delightful, or easy to understand, or perfectly reasonable from the point of view of common sense. The theory of quantum electrodynamics describes Nature as absurd from the point of view of common sense. And it agrees fully with experiment. So I hope you can accept Nature as She is – absurd. (p. 10)

I interpret Feynman's words as supporting the methodological argument that the requirement of prediction is necessary in giving explanations, and without it empirical science is not possible. By contrast, the requirement of an explanation considers, in my opinion, other questions, such as those about causes, factors, and mechanisms, questions which are worth answering, but they are not on the same level of importance as the requirement of empirical prediction. (Basically, these causes and mechanisms increase the ability of the empirical generalizations to generate predictions.)

7.2.2.2 *Transition from descriptive generalization to explanatory generalization*

As stated, Cummins (2000) suggests that in psychology there are no laws similar to those characteristic of the natural sciences, and that in fact these laws are nothing but a kind of phenomena, generalizations, empirical correlations, which require explanation. I believe so too, but with the following differences. First, empirical generalizations, as I said above, do provide us with certain explanatory information; second, some of these generalizations are stable and anchored to firmly based theoretical-empirical knowledge, so they may well be an important tool for explanation of specific behaviors (and see Woodward, 2000, for fixed, stable empirical generalizations, in the special sciences). This theoretical-empirical anchoring is in fact part of the way in which science develops:

(1) First stage: as a result of the research process researchers reach an empirical generalization or a certain law, such as Hook's law, Galileo's law, Kepler's law, and so on.

(2) Second stage: the laws of the first stage acquire broad theoretical-empirical grounding (despite the problematic of this move), for example, grounding of the laws we noted in (1) in Newtonian theory, which provided an explanation for these laws.

Clearly, then, if I want to explain behavior of a body in free fall, I must make use of Galileo's law. But if I want to answer the question, why is the body drawn to earth, I must make use of Newtonian theory of gravity. (Here it is worth noting that the explanation of gravity and impact at distance, an explanation given by Einstein's general theory of relativity, appeared long after Newton.)

Similar processes occur as part of the development of psychology as a science. As an example let us examine the following correlation, which I shall call "aggression displacement": failure (e.g., in finding a job, in personal relations, in exams) arouses aggressive behavior toward people, animals, and objects that were not involved in the generation of the feeling of failure. Now, how may we explain this observation: Ronny failed his entrance exams to psychology and was rude to his parents at dinner. May we not be helped by the "aggression displacement" correlation and say: Ronny's behavior is explained by this correlation, whereby everyone who fails tends to displace aggression; Ronny failed and therefore it is expected that Ronny will respond with aggression displacement. Let us see.

If we take an empirical generalization as no more than a collection or a group of similar cases (i.e., the generalization is nothing but: case 1 plus case 2 plus case 3…) it has no explanatory power for the following reason. Science does not accept an explanation of phenomenon (a) by phenomenon (b) as a proper scientific explanation. If we accepted that a phenomenon explains a phenomenon, we would be able to explain phenomenon (1) – Ronny's aggressive behavior – by phenomenon (2) – Danny's aggressive behavior; and the reverse: phenomenon (2) by phenomenon (1). That is, we explain Danny's behavior as being similar to Ronny's, and we explain Ronny's behavior as being similar to Danny's. The problem is that these explanations infringe a basic property of giving explanations, what I call "direction of the explanation": while the explanans confers understanding onto the explanandum, the explanandum does not confer understanding onto the explanans. (The explanandum supports the hypothesis under study.) For example, while a broken window is explained by the impact on it of a rock weighing one kilogram hurtling through the air at a speed of twenty kilometers per hour, it is difficult to explain what is the source of the rock's motion by an appeal to the phenomenon that the window is broken (here we need additional information, e.g., the ruffian Shimshon hurled the rock at the window).

However, if we take an empirical generalization as *saying something more* that goes beyond being a caption for a group of cases (like what I suggested in the above analysis in which a generalization has a structure of the kind Y = X(2+X)), the generalization is likely to provide us with partial explanatory information. In this case the generalization of aggression displacement as providing information beyond its being a group of cases may well suggest a certain explanation, which depends on the structure of the generalization, for Ronny's aggressive behavior toward his parents. Furthermore, in the present case the direction of explanation is also preserved in the sense that it is hard to posit that Ronny's behavior explains the structure of the empirical generalization of aggression displacement. (As stated, at most Ronny's behavior supports this generalization.)

This argument that the generalization says something more is also underpinned by its fitting into a broad theoretical-empirical framework that provides an explanation for the generalization. As the empirical breadth of the generalization expands, and as its insertion into an established theoretical framework deepens, so its explanatory power increases. If the phenomenon of aggression displacement is generalized over different responses of aggression, different aggression-arousing stimuli and states, over people (sex, culture, race), over the behavior of animals (and see on displacement activity in chapter 5), and if we can provide this generalization with an suitable explanation by means of the multi-explanation theory (because the aggression displacement is a complex behavior that requires mechanistic and mentalistic explanations), the explanatory power of the generalization will be strengthened.

As a further example let us look at the "matching law" that I discussed in chapter 4 (and see Davison & McCarthy, 1988; Herrnstein, 1961). In brief, the law posits that given two responses, response A and response B, where each response is reinforced by a different number of reinforcements (rf), the proportion in which the subject will respond with A, P(A/A+B) exactly equals the proportion in which this response is reinforced $P(rf_A/rf_A+rf_B)$. This law can be applied to explain various learning phenomena: for example, a hungry rat that receives 40 grains of food when it presses lever A and 60 grains of food when it presses lever B learns to press lever B in proportion equal to 60%. The explanation follows the logical structure of the D-N model:

Assumptions: (1) the matching law: the proportion of reinforcements for lever B = the proportion of pressings on lever B,

(2) particular conditions: a hungry rat receives 40 grains of food when it presses lever A and 60 grains of food when it presses lever B.

Conclusion: the observed proportion of pressings on lever B is 60%, therefore the rat's behavior is explained.

As I argued above (and see chapter 4) this law does not meet all the requirements of a scientific law. For example, the law does not meet the requirement of equality of units. The combination of units of measurement of the reinforcement (kind and quantity of food) is not identical with the combination of units of measurement of the response (pressing the lever with the foot). Now, if the matching law is not any sort of law in the wide sense of the natural sciences, how is it to be understood? In my opinion this law is to be understood as a "stable empirical generalization" or "stable correlation": a correlation that is preserved over a large number of subjects, different kinds of subject, different kinds of responses, over different kinds of states, of experiments, and in addition to all these, this empirical generalization fits into and is reducible to the economic theory (see Davison & McCarthy, 1988). In this sense, then, this stable correlation approaches a scientific law: it is likely to support to some degree counterfactuals, to meet to some degree the criterion of universality, and therefore it may be used to explain a given behavior, even though the explanatory power of this correlation, it seems to me, is smaller than the explanatory power of a scientific law as accepted in the natural sciences.

The fact that a given generalization has broad empirical grounding is likely to support various interpretations. We may look, for example, at the genetic-evolutionary interpretation. If indeed a given generalization is sustained over various subjects (humans and animals) and in a large number of empirical situations, a hypothesis may be raised that the generalization apparently expresses a basic behavioral quality, one that perhaps has a foothold in genetic-evolutionary processes. In this case, the empirical generalization acquires the power to induce a genetic-evolutionary explanation for additional empirical phenomena. For example, why did Ronny react aggressively to aggression against him (a powerful blow to his body)? Because by virtue of heredity all humans and animals respond with aggression to aggression. Furthermore, in the present case the direction of explanation is maintained too: it would be hard to suggest that Ronny's behavior explains the genetic-evolutionary theory of aggression – it only supports it empirically.

As stated, the genetic-evolutionary approach is not the only one by whose means an empirical generalization can be moved to the status of explanatory mechanism. Cognitive psychology grounds its explanations in analogy to the computer: the information processing approach. The moment a scientist expresses her correlation in a theoretical framework grounded in a mechanistic mechanism of representations and computation which reflects cognitive processes, at that moment the correlation passes from an epistemological status of description of phenomena that requires explanation to the status of a theoretical-explanatory mechanism, one by whose means we bestow understanding on observations. As an illustration, we may look again at the empirical generalization that the informative capacity and the duration of short-term memory are extremely limited. This is a generalization that has received widespread empirical support and theoretical grounding in the framework of the cognitive approach to information processing (e.g., Baddeley, 1976. And see on this matter chapter 6, where I suggested that the dual theory of memory is in fact constructed on mechanistic and mentalistic explanations, that is, on the multi-explanation theory). Now, is it not possible to use the generalization of short-term memory as an explanation for Danny's self-reproach for forgetting Ruth's telephone number? Listen, Danny says to David, I'm quite senile already. Look, I can no longer remember a seven-digit number for more than half a minute. Please let me have Ruth's number once again. And David reassures Danny, telling him that his behavior is entirely normal, because it accords with the empirical generalization about short-term memory.

The price of solving theoretical grounding by means of analogy with the computer is not cheap. True, in one sense the theoretical grounding inspires explanatory power to the empirical generalization, but in another sense great danger lies here because analogies are vulnerable to replacement. And in psychology, replacement of analogies is an earthquake, a scientific revolution, because this replacement is a replacement of the explanatory mechanism. As an example we shall briefly examine the cognitive and the behaviorist revolutions.

As stated, cognitive psychology is based on the analogy or the metaphor of the computer. The cognitive system (which deals with perception, learning, memory, emotions, thoughts, etc.) is taken as a mechanistic system that processes information, like the computer which takes information in, processes it by means of certain software, and emits the result – the output. What will happen if the metaphor is replaced? To my mind, the moment that psychologists replace the scientific metaphor in which they work with another metaphor, different from its predecessor, a real scientific revolution will take place in psychology. The history of psychology indeed supports this interpretation.

In the early 1960s a revolution befell psychology: behaviorism was replaced by cognitive psychology. This replacement was based on a replacement of metaphors: from a mechanical perception of the individual as acting like a robot, like a machine that responds to a given stimulus with a given response, to the perception of the individual as a computer, as a system that processes information – as representing the world and acting on these representation according to certain rules of computing. This was a veritable revolution, and heads rolled: the cognitivists ignored practically any theoretical and empirical knowledge that the behaviorists had labored so hard to gather and develop in their researches. They simply did not relate either to the great theories of behaviorism or to the impressive harvest of empirical phenomena that were revealed in their studies. The cognitivists began to rewrite psychology: they investigated a large number of phenomena that had no roots in behaviorist research. I know of no such staggering happening in physics. If you're looking for a scientific revolution akin to a political revolution, which annihilates the *ancient régime* in toto, you'll find it here in psychology. Clearly then, in such a situation it is not possible to develop a general theory, as physics did, a theory that can supply explanations for a wide range of observations, simply because with the replacement of the metaphors the theories too are replaced, and with them also a hefty part of the observations that were linked to these theories. For example, the wide-ranging behaviorist theories of Hull and of Spence (see discussion in Kimble, 1961), the theories that dominated psychology in the 1940s and 1950s, were simply tossed aside – they and all the behaviorist phenomena they had succeeded in explaining. Cognitive theories are not based on them, nor do they debate them, or the explanations that they proposed for different and varied experimental findings, so they are not to be seen as theories that made any change whatsoever in earlier concepts.

Here I can't help but recommend to the reader to cease pitying "poor" decapitated behaviorism, because it did to structuralism psychology of the school of Wundt and Tichener (see discussion in Marx & Cronan-Hillix, 1987) precisely what cognitive psychology did to it. Structuralism is a psychological approach that preceded behaviorism, and tried to found psychological research on introspection, that is, on research of the conscious by means of observations of the inner eye. Behaviorism wiped out structuralism, and threw all the observations and the theories, the work of years, straight into the trashcan. Like cognitive psychology, behaviorism began to write psychology anew, and to develop without building itself through discussion with the empirical theories and findings accumulated by structuralism.

Does it emerge from this discussion that psychology is likely to undergo another earthquake in the future, a revolution that will do away with cognitive psychology – the dominant approach today? I think so. If the natural sciences develop a new system in the future, different from the present computational system, a system that can provide psychology with a new metaphor by means of which it will be possible to construct a new mechanistic explanation, we will be witnesses to a new revolution in psychology (see Rakover, 1990, 1992).

The question that arises in consequence of this discussion is the following: were the scientific revolutions in psychology justified? If indeed structuralism or behaviorism created a poor and trivial science, these revolutions were justified. If not, beheading was not condign, and the replacement of metaphors there seems several times more cruel. I believe that at least regarding behaviorism the decapitations were a misplaced "orgy". In 1986 I wrote an article criticizing a paper by the philosopher of science Suppe (1984) about the scientific development of psychology. In this paper he raised the argument that behaviorism was a trivial, poor and inferior science. In my article in response I showed by means of a logical analysis and an analysis of various behavioral phenomena that behaviorism had revealed, that the situation was the diametrical opposite: behaviorism created a science of extreme interest, which in many respects did not differ from the science created by cognitive psychology. Sixteen years later I chanced to read an article by Hinzman (1993) in which he compared behaviorism and cognitive psychology in the empirical and theoretical respects, and he reached a conclusion similar to mine. The title of Hinzman's article says it all: "Twenty-five years of learning and memory: Was the Cognitive revolution a mistake?"

CHAPTER 8

Establishing multi-explanation theory (b)

Methodological dualism

The present chapter aims to suggest a theoretical basis for multi-explanation theory by setting forth reasons why no attempted solution to the mind/body problem has yet succeeded. Were it possible to reduce consciousness to the neurophysiology of the brain, a multi-explanation theory would have no justification because all would be explained in the framework of concepts of the natural sciences. The chapter examines five main research areas connected to the mind/body problem: mental causality, functionalism and the argument of multiple realizability, the computer and the process of decomposition into basic mechanisms, neuropsychological reduction, and consciousness; it concludes that indeed no philosophical or scientific approach has yet been found that offers an acceptable solution to our problem. Hence there is room for the approach of methodological dualism, whereby complex behavior of humans and animals must be investigated on the basis of a combination of mechanistic and mentalistic explanation schemes. Furthermore, in this framework the chapter raises several theoretical and empirical reasons showing that consciousness is not an epiphenomenon but an important and essential factor in the explanation of behavior.

An interstellar trip is absolutely safe because it is guided by super computers. But I skipped the trip because on the long and tedious journey these computers suffer attacks of deep clinical depression when they aren't able to solve the mind/body problem.

In the last chapter I tried to establish the multi-explanation theory on the argument that mentalistic explanation schemes are an important element in the methodology of the social sciences, even though they differ in several respects from explanation schemes followed in the natural sciences. In this chapter I discuss the argument that I partly considered in previous chapters that mentalistic explanations need to be proposed for the behavior of animals, principally because mechanistic theories have not been able to offer complete explanations for complex behavioral phenomena; because no way has yet been found to understand the mind/body problem, the connection between the brain and consciousness, that is, to find the place of the phenomenon of consciousness in the framework of science. (Here it is appropriate to comment that I refer to the mind/body problem broadly, that is, as a problem closely bound up with a plethora of questions, for example, mental causality, reduction of the mind to the body, and consciousness.)

If it were possible to understand the existence of consciousness in terms of the natural sciences, the methodological basis for a multi-explanation theory would be undermined. I call this basis "methodological dualism" (it proposes a methodology for constructing theories that utilize mechanistic and mentalistic explanations equally), because not everything can be explained by means of the conceptual framework developed in the natural sciences.

Similar things were written by McCauley & Bechtel (2001) in the context of reducing a psychological theory to a neurophysiological theory according to the classic model of reduction (see below):

> … if psychological theories map neatly on to neuroscientific theories along the lines that classical model specifies, then our commitments to psychological states and events are, at very least, dispensable in principle. (p. 739)

It does not follow from this that if no solution to the mind/body problem is found methodological dualism has to be adopted. My argument is weaker: this situation of unsuccessful attempts to offer a solution to our problem leaves room for methodological dualism. Nevertheless, ongoing research may well develop new methods, new methodologies, for tackling our problem, and maybe one day science will make a breakthrough so great that our understanding of the mind/body problem will be realized. When will this happen? I do not know.

Methodological dualism, which in essence is not ontological dualism (see Robinson 2003), proposes, on the one hand, that consciousness is an important and basic part of the behavior of humans and animals, and on the other hand that the mentalistic explanation has methodological autonomy principally because, as stated above, no way has yet been found to explain or to reduce the mental to the mechanistic. (Methodological dualism is close in spirit to a soft kind of dualism, to "explanatory dualism", which suggests that the explanation of natural phenomena is essentially different from the explanation of mental actions and behavior (see, e.g., Brook & Stainton, 2001; Maxwell, 2000; Sayre-McCord, 1989). I shall discuss the comparison between methodological dualism and the explanatory dualism in chapter 9.)

No philosophical approach exists showing that it is possible to understand the experience of consciousness in material terms; and no scientific theory exists showing how a mental phenomenon is composed of material elements, or how mental processes turn into neurophysiological processes and the reverse, just as in natural science the composition of materials and transformation of energy from one kind to another are explained.

I believe that the mind/body problem can be expressed by means of three stances, well anchored to everyday knowledge and to accumulated scientific knowledge:

(A) *The everyday knowledge stance*: for the individual, subjective conscious phenomena, such as feelings, emotions, thoughts, images, desires, and beliefs, are as real as phenomena of the physical world.

(B) *The modern science stance*: physical, chemical, physiological, and behavioral phenomena (such as motor movement), that is, phenomena contained in the material domain, are real and are explained by science mechanistically.

(C) *The joint stance*: conscious subjective phenomena and physical phenomena are perceived as natural phenomena and as an integral part of the world.

From these stances, to my mind, three hypotheses arise, for which I shall try to find justifications:

(a) *Mind/body connection*: although we dispose of no theory explaining the nature of mind/body relation, we perceive this relation as obviously real.

(b) *Consciousness as an explanatory factor*: consciousness is an important component in the explanation of actions performed intentionally, rule-following, feelings and emotions such as pain, fear, anger, frustration, insult, joy, and laughter.

(c) *Consciousness as a phenomenon requiring explanation*: consciousness as a phenomenon and as an essential component in complex behavioral phenomena requires an explanation by an appeal, among other things, to cognitive and neurophysiological processes.

I shall justify these three hypotheses by anchoring them to examples. I shall begin with (a) *mind/body connection*. By virtue of scientific knowledge that has become public domain we know that not all actions that take place on our body (such as chemical and neurophysiological processes) are connected to mental processes, and that we are not aware of all the processes that occur in our body. However, we are aware that a large number of cases exist that express the mind/body connection, for example, performance of planned activity and (different) feelings of pain from a blow, a prick, a burn. These cases powerfully support the hypothesis that we (and other animals) are indeed creatures in which a not understood but distinct connection of mind/body takes place.

Hypothesis (b). *Consciousness as an explanatory factor* is expressed in the following examples. If the mental events according to which David wishes to meet Ruth, and believes that a journey to Tel Aviv will realize this wish, are not located in his consciousness, in David's awareness (e.g., because he has forgotten them entirely), his journey for the meeting with Ruth will not materialize. For David, the wish/belief and the action that is planned to realize his intention do not exist when these are not present in his consciousness. David can travel to Tel Aviv for other reasons, but not because of his wish to meet Ruth. Similar things may be said about rule-following, emotions, and the like. Furthermore, I argue that without subjective consciousness of each and every person, cultural heritage would not be possible. Rakover (1990) distinguished two kinds of mental states. Mental state$_1$ (MS_1) refers to sensations, feelings, emotions, etc., with which the individual is endowed from birth, and which are expressible and reportable publicly (partly) by means of mainly verbal behavior. MS_2 relates to what is public, for example, language, knowledge, values, which the individual internalizes by a learning process, which of course she can also express publicly. In this respect, if we do not hand down our cultural heritage from generation to generation, namely if we do not take care to instill this heritage into the consciousness of people, into the consciousness of each and every one, the culture will not be bequeathed to the next generations; it will be erased, and will be nothing but strange, unintelligible marks on yellowing leaves of a book. That is, consciousness is not only essential for intentional action and for other behavioral phenomena, it is essential for the very existence of culture.

Hypothesis (c). *Consciousness as a phenomenon requiring explanation* is expressed in the following examples. While a gash on the hand arouses a feeling of pain in me, this feeling is eliminated by local anesthesia. A change in the conscious state (e.g., by diverting attention, hypnosis) will cause elimination of the feeling of pain. In many cases, such as at times of great excitement, people are not aware that they have been wounded and discover that their leg hurts afterwards. Consciousness varies not only in its strength, but also in its kind (e.g., according to feeling, to the strength of kinds of the physical stimulus). Moreover, we are immersed in a continuous existence of consciousness consisting of a collection of events simultaneously present in our conscious. For example, I am now concentrating on writing this line, but I am also aware of the computer screen on which this line appears, of the sound it makes, of the feel and click of the keys, of the light in my office, of the classical music coming from the radio to my left (Aaron Copland's *Billy the Kid*), and so on.

Now, given these stances and hypotheses, how is it possible to support the claim that no way has as yet been found to understand the mind/body connection in the framework of the concepts of modern science? I believe that it is hard to find proof showing it is impossible to find a connection between mental state (MS) and neurophysiological state (NS) similar to a logical or a mathematical proof, because *the mind/body problem arises, to my mind, in the framework of the game-rules of empirical science based on hypotheses and observations, and in this framework it is to be solved.* My impression is that the solutions offered to the mind/body problem encounter several basic obstacles again and again, which end in a cul-de-sac.

(1) *Epiphenomenalism*: Some of the important relations between MS and NS, which were developed by modern philosophy in an attempt to explain MS in terms of NS, focus on the relation of supervenience, which proposes that a change in MS is not possible without a change in NS, and on the relation of realizability, which proposes that MS is realized by NS. These relations create a dilemma, I think. MS has no explanatory function, so the status of MS proves irrelevant, epiphenomenal (MS is influenced by NS but cannot influence NS or behavior), because NS does all the explanatory-causal work; on the other hand, everyday intuition tells us that consciousness, MS, has an evident explanatory role in the behavior of animals (see above, especially (a) and (b)). Kim (1966), discussing these relations with an example from pain, the contraction response, and the neural state (N), writes:

> It is incoherent to think that the pain somehow directly, without an intervening chain of physiological processes, acted on certain muscles, causing them to contract; that would be telekinesis, a strange form of action at distance! ... It is implausible in the extreme to think that there might be two independent causal paths here, one from N to the wincing and the other from the pain to the wincing. (p.150)

(2) The "*qualitative gap*": It seems that it is very hard to understand how neurophysiological states and processes create or become conscious, a conscious experience. A qualitative gap opens here between the above two stances: (A) the everyday knowledge stance and (B) the modern science stance.

(3) "*Empirical measurement*": While in science, theories are constructed that describe-explain mechanistically the connection between one NS and another, and between neurophysiological and motor behavior, no theory has been found that is able to describe-explain mechanistically the connection between NS and MS or between MS and behavior. The reason lies in the non-ability to measure conscious phenomena, MSs, as we measure physical and neurophysiological events (see above and discussion in chapter 4);

(4) "*Empirical testing*": The structure of the hypotheses, the theories, proposed as a solution to the mind/body problem in most cases impedes their empirical testing, and turns them largely into ad hoc explanations.

(5) "*Causal explanation*": In a large number of explanations of the mind/body connection no account is taken of the fact that the causal explanation does not uphold the rule of transitivity (I shall call this feature "non-transitivity"). From the two assumptions (a) is bigger than (b) and (b) is bigger than (c), it necessarily follows that (a) is bigger than (c); but it does not follow from the two assumptions (a) causes (b) and (b) causes (c) that (a) causes (c). For example, from the fact that squeezing the trigger causes the primer to strike the propellant in the cartridge which explodes and causes the bullet to fly, it does not follow that squeezing the trigger in itself causes the bullet to fly. Owens (1989) suggests that the causal explanation is not transitive because the chain of reasons can come about accidentally, for example, in the case where person A finds himself by chance in a certain place, and there he is murdered by person B. By contrast, the reason for non-transitivity that I suggest is based on empirical testing: does squeezing the trigger in itself cause the flight of the bullet? Is (a) indeed bigger than (c)? (Owens notes that even Hempel (1965) tends to agree that the causal explanation does not exhibit a relation of transitivity.) Hence from the proposal that neurophysiological processes arouse consciousness and consciousness affects behavior, it does not necessarily follow that neurophysiological processes in themselves cause behavior, because according to the property of non-transitivity there is room for consciousness as a causal factor for behavior (and see on this question Flanagan, 1992; Rakover, 1996; Velmans, 1991).

An important variation on the non-transitivity argument is what I call the "eliminative argument" (see Rakover, 1983b, 1990). Here we assume the following relations, or functions:

(a) Response = f(Mental state)

(b) Mental state = g(Stimulus)

From (a) and (b) follows (by mathematical substitution) (c) Response = fg(Stimulus), so as may be seen, the mental state does not appear in (c).

This argument is erroneous for two principal reasons. First logical relations cannot be used to infer ontological existence or nonexistence (of an entity). For example, from the fact that motor revolution in a car depends on pressure on the pedal, and from the fact that speed of travel of the car depends on the revolution of the motor, it does not follow that the speed of the car depends on pressure on the pedal alone, and that it is possible eliminate the action of the motor in explaining the action of the car. Secondly, in the complex function in (c) – fg, the functions (a) and (b) actually do appear.

Assuming that the different approaches to the mind/body problem have yet to yield solutions, the following question arises: what is the "solution-limit" that present-day scientific methodology can reach in the attempt to explain this problem? To my mind, the solution stops at the methodological limit of the correlation between indices expressing the mental and indices expressing the behavioral-neurophysiological. For example, from David's report of his *wish* to meet Ruth in Tel Aviv and from the report on his *belief* that driving to Tel Aviv in his car will bring his wish to fruition, we conclude in practice that he indeed will carry out his intention and will *drive* to Tel Aviv in his car; from perusal of the behavioral changes as a result of the formation of brain lesions, or as a result of a brain stimulation by an electric current, it is possible to construct interesting correlations between neurophysiological activity in the brain and various indices of consciousness and behavior; and from looking at functional magnetic resonance imaging (fMIR) it is possible to draw up a correlation between, for example, areas of activation in the brain and a feeling of pain. But note here that even a correlation of the latter kind is problematic: Uttal (2001) in his book *The new phrenology* sharply criticizes the idea of local correlations in the brain, namely hypotheses concerning brain localization of mental processes by use of modern imaging techniques, for such reasons as the difficulty of defining and decomposing mental and neurophysiological processes to basic distinct units, and the enormous complexity of the brain and of cognitive processes.

In this chapter I shall review several subjects relevant to our concern, and I shall show that the road to understanding the mind/body problem is indeed long. A number of researchers have stated this conclusion expressly:

1) Whatever our mental functioning may be, there seems to be no serious reason to believe that it is explainable by our physics and chemistry. (Putnam, 1975, p. 297)

2) We have been trying for a long time to solve the mind-body problem. It has stubbornly resisted our best efforts. The mystery persists. I think the time has come to admit candidly that we cannot resolve the mystery. (McGinn, 1989, p. 349)
3) The position that I am elaborating here, however, suggests that neither behaviorism nor dualism – in their classic or contemporary guises – can solve the problem of nature of the mind, which requires an alternative approach that regards behavior as evidence for the existence of non-reductive, semiotic modes of mentality. (Fetzer, 2001, pp. 3-4)
4) The reason the mind-body problem does not go away, despite our being clear about the options in responding to it, is because of the constant battle between common sense, which favors the view that the mental is a basic feature of reality, and the pull to see it as an authoritative deliverance of science that this is not so. We find ourselves constantly pulled between these two poles, unable to see our minds as nothing over and above the physical, unwilling to see the universe as containing anything not explicable in terms of its basic, apparently non-mental, constituents. (Ludwig, 2003, pp. 29-31)
5) Even if we accept the familiar idea that minds are somehow dependent on brains, we have no clear idea of the nature of this dependence. The mental-physical relation appears utterly mysterious. (Heil, 2003, p. 217)
6) The problem of consciousness is completely intractable. We will never understand consciousness in the deeply satisfying way we've come to expect from our sciences. (Dietrich & Hardcastle, 2005, p. 1, opening sentence)

These views, of course, are liable to ignite debate. For example, Flanagan (1992) does not agree with McGinn's pessimistic argument and conclusion. According to McGinn, since on the one hand observation on the brain allows conclusions only about interactions between neurophysiological events, and since on the other hand internal observation, introspection, allows conclusions to be drawn only about connections between mental events, the connection between these two observations will remain a sort of mystery. By contrast, Flanagan suggests that this conclusion is too extreme, and does not entirely remove the possibility of proposing a theory according to which NSs will function as realizations of MSs.

Now I shall move on to discuss very briefly five main areas of research on a solution to the mind/body problem: mental causality; functionalism and multiple realizability; the computer and the decomposition process; reduction; and consciousness. The aim is to test if indeed they have succeeded in proposing solutions to our problem in these areas. My impression is that we are still far from resolving this mystery, so to my mind, room is created for the methodological dualism approach.

8.1 Mental causality

Why do we believe that an appeal to the individual's will and belief constitutes an explanation for her actions? This is one of the basic questions of the philosophy of mind/body, the philosophy of action, of practical reasoning, and of psychology itself (e.g., see discussions in Graham, 2002; Kim, 1996, 1998; Mele, 2003; Millgram, 1997; Rakover, 1990; Rosenberg, 1988). One of the important solutions to this question of ours is the proposal to regard the teleological, will/belief explanation, as a causal explanation. Davidson (1980) suggested in the article "Mental events" that a true explanation given by means of reasons, that is, an explanation through an appeal to the individual's desire and belief, is fundamentally a causal explanation (and see discussion on this subject in Kim, 1993). In his view, only thus is it possible to answer the question which out of several reasons is the one responsible for the investigated behavior: it is the reason that is a physical event that constitutes the cause of the given behavior. His suggestion is based on a solution to the lack of consistency of the following three principles:

1) The principle of causal interaction: This principle proposes that a number of mental events are in causal interaction with physical events. (This principle is parallel to the hypothesis about the mind/body connection that I formulated above.)
2) The principle of the nomological character of causality: This principle suggests that if a certain causal relation exists between two events, this relation is covered, explained by a determinist scientific law that connects the event of the cause to the event of the outcome.
3) The principle of the anomalism of the mental: This anomalism suggests that no determinist law exists by means of which mental events can be predicted and explained. (Davidson bases this anomalism chiefly on the fact that a psychophysical "law" is not possible, because this "law" is based on a person being rational and consistent, and not on a mechanistic causal connection. This principle is parallel to the position of everyday knowledge and to the stance of modern science that I formulated above.)

However, although each principle in itself is perceived in our eyes as extremely convincing, the three principles together do not create unity or a group of principles free of internal contradictions. From the first two principles together (causal interaction between mental and physical events, and the existence of a causal law) the conclusion arises that some mental events are predictable and explicable by physical laws. But the third principle – mental anomalism – contradicts this conclusion.

Davidson's idea is to suggest a solution to this inconsistency, to show that the inconsistency is an illusion. In his opinion, what harmonizes these three principles is the proposal to regard every mental event as a physical event. This identity is perceived not as identity between different kinds, types, of mental and physical events (called mind-body type identity theory), an identity which in fact constitutes a psychophysical law which according to Davidson is not possible, but identity of elements, of individual events, of mental "particulars" and physical "particulars" (called mind-body token identity theory).

Hence the connection between mental events and physical events is causal, because in the end specific mental events are specific physical events placed in causal interaction between themselves. Davidson called this solution anomalous monism, because the mental event is a physical event, and because the third principle is upheld by this solution.

Does anomalous monism offer a solution to our question? Does the solution lie in the suggestion that the explanation for the individual's action through her desire and belief is actually a causal explanation that assumes identity between specific mental and physical events? As may be expected, Davidson's thesis aroused great criticism and interesting philosophical debates, which I cannot go into here (and see Kim, 1993). Nevertheless I would like to point out three problems.

First, if the explanation of the connection between two given events is accomplished entirely on the basis of a causal law between physical events, because a mental event is in fact a physical thing, where is the location place of the mental event in itself in the explanation of the individual's behavior? According to Davidson, it transpires that all the explanatory work is set on the shoulders of the physical-causal law, and therefore mental properties, as mental properties, have no explanatory status: the mental is epiphenomenal. This conclusion runs counter to our strong everyday intuition that in most cases people act according to will/belief, according to their mental states. This argument, evidently, is a kind of variation on the argument of epiphenomenalism that I formulated above, and if it indeed holds it transpires that according to Davidson's approach there is no need for psychological theory, because the prediction and the explanation in any case are accomplished wholly by the natural sciences, that is, by NSs.

The epiphenomenalism obstacle is apparently an insurmountable hurdle. For example, Kim (2002) writes in his précis for his 1998 book *Mind in a physical world* the following final conclusion:

> To summarize, then, the problem of mental causation is solvable for cognitive/intentional mental properties. But it is not solvable for the qualitative or phenomenal characters of conscious experience. We are therefore left without an explanation of how qualia can be causally efficacious; perhaps, we must learn to live with qualia epiphenomenalism. (p. 643)

The epiphenomenalism obstacle is a counter-argument to functionalism (see below) according to which a mental state may be characterized by means of its function, the functional property, that this state has in a given system. The argument is that the functional property does not possess explanatory power in itself, and that the explanation is made by the neurophysiological state that realizes the function of the mental state; that the functional description in fact is nothing but a kind of description of a function that is to be explained by means of material (neurophysiological) processes (see discussion in Looren de Jong, 2003, who believes that functional explanations are valid). This is the place to note briefly that arguments of this kind bring to mind the instruction proposed by Kant (1790/1964) in his book *The critique of judgment* that teleological explanations have a heuristic, useful value, which helps in understanding the world, so they may be used until the true causal explanation is found.

Secondly, in my opinion the connection between mental and physical events, characterized as token identity, is not found in science's area of interest. Science is not concerned with the particular, individual connection between event X and event Y, but with that between certain kinds of Xs and certain kinds of Ys. Why? Because it is not possible to test a specific empirical connection, to predict or suggest an explanation for a specific observation, if all we know about this observation is that it is connected to another specific observation (hence an ad hoc connection). To suggest an explanation for X and Y we have to show that these X and Y belong to some empirical law. But Davidson's thesis denies the possibility of finding such a law (and see argument above: empirical testing).

Thirdly, the proposal of token identity, that $MS_1 = NS_1$, is not parallel to the scientific discoveries of identities, for example, water = H_2O, which deal with kinds of identities. While the latter proposition is explained by and anchored to a research process, the first proposal is nothing but a kind of metaphysical possibility, which may perhaps extricate us from several philosophical problems concering the relation between consciousness and the brain. The identity water = H_2O can be explained by an appeal to chemical theory, which, for example, may be tested empirically in the following simple way: if we introduce into a container quantity X of hydrogen and quantity Y of oxygen (in atomic proportion 2:1) we will obtain water in the amount X+Y. And the reverse: by means of an electrical process called electrolysis water may be broken down into its components, hydrogen and the oxygen. No similar explanation can be suggested for the identity $MS_1 = NS_1$. To the best of my knowledge there is not a single empirical theory that explains this identity (not how it is possible to obtain a mental state from neurophysiological components, and not how to break down a mental state into its neurophysiological components), among other things because it is not possible to measure MS_1 as one measures NS_1. All we can achieve is no more than a correlation, an association, between mental states and neurophysiological states (and see arguments above: measurement and empirical testing).

Finally it is worth noting that while I stress the importance of the scientific research process, which in certain cases also reveals and explains identities such as the water/gases identity, Kripke (1972/1980) suggests that this identity is a necessary truth a priori, and not a posteriori, that is, a truth that was discovered at the end of the scientific research. This argument comes out against the theory of identity of types, and Polger (2004) debates it as part of his theoretical attempt to breathe fresh life into the theory of identity of types.

8.2 Functionalism and multiple realizability

Functionalism proposes a different solution to the mind/body problem, one that since the 1960s has in one way or another been a philosophical basis for cognitive psychology, a solution that developed, among other things, as an alternative to identity theory (see discussion in Heil, 1998; Lycan, 2003; Rakover, 1990). The functionalism approach has many different variations. For example, Polger (2004) argues that according to his method of classification there are over 100 variations of functionalism, but he discusses six kinds thoroughly (Churchland, 1988, by contrast, deals with four kinds of functionalism). Common to these variations is the attempt to understand a mental state from the viewpoint of its function, the role that this state plays in the system it belongs to. What distinguishes them is the sort of answer offered in reply to the central question of the given variation. For example, metaphysical functionalism focuses on the question of the nature of the mental state; intentional functionalism focuses on the question of the 'about', the 'intentionality', of the mental state; explanatory functionalism focuses on the question of the explanatory role of the mental state; and so on.

I shall not enter into a discussion of these matters, but only comment that generally it is possible to characterize functionalism as a theoretical approach according to which a mental state is defined by a functional property, which may be described as connecting the stimulus to the response, as interacting with other states, and as realized materially (e.g., by the brain). Kim (1996) writes:

> ...psychological concepts are like concepts of artifacts. For example, the concept of 'engine' is silent on the actual mechanism that realizes it – whether it uses gasoline or electricity or steam ... As long as a physical device is capable of performing a certain specified job, namely, that of transforming various forms of energy into mechanical force or motion, it counts as an engine. The concept of engine is specified by a job description of mechanisms that can execute the job. (p. 75)

This notion does not accord with type identity theory, which proposes that a given type of mental state is identical to a certain type of cerebral state, mainly because according to the functionalist approach a mental state can be realized by different types of materials. Functionalism, then, raises one of the most powerful arguments against the possibility of suggesting a psychophysical law according to which the type of mental state is identical with the type of neurophysiological state, just as identity exists between lightning and electrical discharge, or between water and a mixture of gases (H_2O), an argument called "multiple realizability". The idea underlying this argument is this: just as the computer program may be run, realized, by different types of computers constructed of different materials, so consciousness is likely to be realized by different materials. And just as a certain computing situation in the program is characterized by connections to other computing situations, to input and output, so a mental state is defined through its relations to other mental situations, to stimuli and responses. According to the functionalist approach, then, a mental state is defined through the function that it serves in the cognitive system. And just as the function "to bid farewell to Ruth" may well be realized through a large number of means (saying goodbye, writing a parting letter, waving), so may a certain mental state, pain, be realized through different materials and in different forms (e.g., by the brain of the human, the monkey, the dog, and the cat, and also by a brain made of silicon). That is, a mental state in X may be equal to a mental state in Y, as long as the arrangement of the (different) realizing materials in both cases is equal in the functionalist respect. The argument can therefore be raised that a realizing material, such as silicon, arranged functionally in a way equal to the neurophysiology of the mental state of fear in humans, will also be the basis of the state of fear. If the argument of multiple realizability is right the theory of consciousness-brain identity is wrong, because the theory of identity will be hard pressed to explain, for example, pain in a dog and a cat (see discussions in Bickle, 1998; Fodor, 1974, 1988; Heil, 1998; Kim, 1996; Polger, 2004; Putnam, 1967).

The functionalist approach has several serious problems due largely to the fact that in essence this approach suggests a mechanistic explanation for behavior. That is, the role of the causal explanation is ascribed not to mental states but to material states, to materials that realize the mental states, for example, to the neurophysiology of the brain. This ascription, as I noted above, runs counter to common, everyday, sense, which suggests that mental states have content, meaning; that they appeal to a certain subject, that they are located in us in consciousness, in awareness, and that they are to a great extent responsible for our behavior. (See criticism of functionalism in Block, 1978; Polger, 2004. The latter, for example, mounts a sharp criticism of functionalism, which leads him ultimately to renewed support for the approach of identity of types of mind/body, which at the time, as noted, were assailed and replaced by functionalism as the theory for the mind/body problem.) Here I shall examine two critiques of functionalism and multiple realizability.

(a) The Chinese Room: If we produced a computer of extraordinary perfection, one whose actions, in a functionalist respect, were structured identically to the way the actions of human consciousness are structured, by the functionalist approach this superior computer should have consciousness like a human. Is this argument convincing? Searle (1980), and I following him (Rakover, 1999), believe that this argument is incorrect. In his famous 1980 article Searle presented his thought experiment, called "the Chinese Room". The essence of the experiment is that Searle undertakes to perform all the actions that the computer does, when this machine processes the Chinese language. Assume, then, that Searle, who hasn't the slightest knowledge of Chinese, assumes the function of the computer, enters the Chinese Room, which constitutes an analogy to the computer, obtains input in Chinese characters, processes them according to books of instruction and guidance in English, does all the computing actions that the computer does with this input, and produces output in the Chinese language. Specialists in Chinese study this output, and determine that only an intelligent person, who knows Chinese well, would be able to answer all their questions (the input is in

Chinese). This last fact shows explicitly that Searle has passed the Turing (1950) test, namely: if we as judges are unable to distinguish between a human and a computer by virtue of the answers that these two give to our questions, it will not be justified to attribute to the computer non-consciousness. That is, if the computer passes the Turing test, we have sufficient justification to attribute mentality to it. (See summary and discussion on the Turing test in Saygin, Cicekli & Akman, 2000.) This test, then, supports the hypothesis that Searle understands Chinese just as well as an expert in that language. It may be said, therefore, that since Searle acted exactly like the computer, the computer too understands Chinese. But Searle, who is human and endowed with a mind, consciousness, and awareness, like you and me, argues the opposite. He claims that he leaves the Chinese Room with the same level of ignorance of comprehension of Chinese as when he entered it. That is, the hypothesis of functionalism, that a highly sophisticated and perfected computer will develop conscious awareness, has failed.

This thought experiment, as may be expected, sparked fierce criticisms. I was not convinced by them, and I gave expression to this in a thought experiment that is a variation on Searle's and supports the criticism on the functionalist approach resulting from his experiment. In my variation I started from the viewpoint of a super-computer that does a simulation on Searle in the Chinese Room. The analysis from this viewpoint showed that the computer is indeed nothing but a machine devoid of consciousness. (See Rakover, 1999. The article also briefly summarizes the controversy on the subject.)

The Chinese Room argument touches on important subjects closely linked to functionalism, and to the Turing test. This thought experiment may be seen as support for the hypothesis that syntactical processes, which the computer accomplishes, do not invest words and sentences with semantic meaning – otherwise Searle would have emerged from the Chinese Room with a good understanding of Chinese. So the question that arises is: how is meaning induced into the products of the computations that the brain performs on physical representations, that is, on the neurophysiological units that realize representations of the individual's environment? This is a difficult question – very broad and ponderous, which aroused many debates that I cannot deal with and make all the appropriate distinctions here, so I shall merely mention some relevant allusions. (It is worth mentioning here that Fodor's (1994) main goal was to reconcile the computational with the intentional. This however, as can be seen from the preface and the previous chapters, is not the purpose of the book.)

According to the "symbol grounding problem" coined by Harnad (1990), it is not clear how it is possible that semantic meaning was imparted to a syntactical system of physical symbols without the process of investment of meaning taking place in a human person's head. In other words, the question is, can Searle, or anyone else who does not live in China, does not belong to the Chinese culture, and does not understand Chinese at all, learn Chinese with all the meanings of that language, using a Chinese/Chinese dictionary alone? Several solutions have been proposed for this problem and others connected to it.

In the framework of the functionalist approach and the perception of consciousness as a process of "symbol manipulation" or "information processing" (also called the "classic approach"), Fodor (1976) put forward the hypothesis on "the language of thought". Accordingly, thoughts are related to mental representations, which are viewed as linguistic expressions created by a basic innate language similar (but not identical) to human language. The most important alternative to the approach to consciousness as a process of symbol manipulation is the connectionist approach, the neural network approach. In this framework computer programs were developed imitating patterns of activation of neural networks in the brain as realizing mental-cognitive processes. Which of these two computational approaches (constituting two different groups of computational models of cognitive systems) is more efficient? This question too gave rise to a debate that is still not over. For example, Marcus (2001), who comes from the field of research on language and higher cognitive processes, compared these two approaches and concluded that a relatively straightforward software of the connectionist approach (simple multilayer perceptrons) finds it hard to tackle several behavioral problems that classic symbol manipulating manages successfully; that is, the neuron network approach does not display adequate power and flexibility. He also suggested that the symbol-manipulating mechanism is innate and developed by evolution. (Interesting discussions on these and other matters may be found, for example, in Block, 1995; Crane, 1995; Fetzer, 2001; Fodor & Pylyshin, 1988; Heil, 1998; Smolensky, 1988).

(b) Zombie and "Robocat": If a physical copy of a human being could be constructed, a double without consciousness, a zombie, functionalism would be incorrect. For example, assume that it is possible to construct such an odd kind of creature, that walks about, that fulfills, for example, all the functions of pain (it screams, curses, weeps, clutches the part that hurts), but without its having consciousness, or any conscious feeling, any experience of pain. If constructing such a creature were possible and conceivable, its very existence would contradict functionalist theory, which holds that this zombie is endowed with conscious existence. This thought experiment, as may be imagined, generated much debate. (See discussions in Crane, 2003; Dennett, 1991; Heil, 1998; Polger, 2004. The philosophical story, of course, is much more complicated. See, for example, discussions on thought experiments of an individual and her double, and on the question of whether the content of mental states originates in the relationship of the brain to the external world (externalism) or whether this content is present in the consciousness itself (internalism): Heil, 1988; Von Eckhardt, 1993.) I shall not enter into this thicket here, but instead make just the following comments.

The use of a zombie as a thought experiment constitutes counter-evidence to the claim that there is no need for mentalistic explanations to understand the complexity of the individual's behavior, and that all may be explained by means of mechanistic explanations. This argument accords with what Flanagan (1992) calls "conscious inessentialism". This is

> the view that for any activity i performed in any cognitive domain d, even if *we* do i consciously, i can in principle be done nonconsciously. (p. 129)

If the possibility that creation of a zombie or a robocat (or a zombicat, for example, a consciousness-less physical copy of Max the cat) is conceivable, the rug is pulled from under the feet of the mechanistic explanation itself. The reason is this: if on the basis of the mechanistic explanation (e.g., functionalism), which aspires to understand behavior in its entirety – *including conscious existence,* it is possible to create a zombicat, a creature without consciousness, it follows that consciousness does not necessarily stem from a mechanistic mechanism and the mechanistic explanation does not have the power to offer an explanation of all behavior. Hence the approach of conscious inessentialism is unfounded. Flanagan does not accept this approach either, and shows by an in-depth analysis of several experiments and clinical cases of brain damage that consciousness has a causal status and is not a kind of epiphenomenalism (and see above on this matter, epiphenomenalism and the causal explanation).

One further point. Let us assume that we create a physical double of Max the cat in a given time, and let us assume that these two live separately in exactly the same environmental conditions for two years. After this time, will the double behave like Max? If your, the reader's, answer is no, you must accept the possibility that consciousness has an important causal status. (I leave this question on the intuitive level, because I believe that thought experiments of this kind are constructed out of an elaborate mix of scientific knowledge, analogies, and powerful personal intuitions.)

8.3 The computer and the process of decomposition

Functionalism, which was proposed as a solution to the mind/body problem, forms the philosophical basis of cognitive psychology and the cognitive sciences. According to this infrastructure, the computer, as a machine that performs manipulations on physical symbols, constitutes an extremely important analogy for understanding psychological mechanisms for processing information such as perception, attention, learning, memory, decision-making, etc. (see Bem & Looren de-Jong, 1997; Copeland, 1993; Franklin, 1995; Haberlandt, 1997; Pylyshyn, 1984; Rakover, 1990; Thagard, 1996; Von Eckhardt, 1993). Within this approach, I shall discuss the following question: does the computer analogy help us explain the relationship between consciousness and the brain?

To answer this question I shall very briefly review how the computer makes computations, answers questions, prints text, and draws figures (see, e.g., Block, 1995; Deitel & Deitel, 1985; Von Eckhardt, 1993). The important point that I would like to start with is that the computer is a mechanism composed of a large number of parts and processes that mediate between the information put into it and the information coming out of it. All these parts and processes are without exception mechanical mechanisms, whose functioning is explained by means of mechanistic models. For our purposes, we shall examine how a software is translated to a series of physical action instructions, which are performed by the computer according to the laws of electricity.

First, the software, which the programmer wrote in a higher computer language, automatically undergoes translations from language to language until it is translated entirely into a language based on several simple actions that the computer is able to perform, logical actions done on chains of zeros and ones (0 and 1), namely on binary digits. Let us look, for example, at how the computer adds up numbers. This addition is carried out by a complex combination of three electrical units that realize the following three logical units:

And is a unit that takes two inputs (each input being able to obtain a value of 0 or 1) and produces 1 on condition that the two inputs are 1. As for any other combination – 00 01 10 – the *and* produces 0.

Or is a unit that takes two inputs and produces 1 where at least one of the two inputs is 1.

Not is a unit that takes one input and changes it from1 to 0 or from 0 to 1.

Without going into the organization of these units, into a diagram of the mechanism of adding up numbers, I shall say only that the mechanism can add any number we like by translation of the decimal language into binary language, and back into the decimal language which we know well. The point I am stressing in this example is this: the computer is able to perform only these three simple actions. When an instruction is inserted into it to add up two numbers it translates them into binary language, into series of 0 and 1, inserts these series into the adding mechanism, performs these simple actions (and, or, not), and in the end yields the desired result.

Secondly, the reason why the computer successfully performs what is written in the software, the above three actions that recurrently appear according to the diagram in the mechanism for adding up numbers, lies in the connection made between the digits 0 and 1, and between two electrical states in the computer: without electrical voltage and with electrical voltage. As a result of this connection the computer carries out physically (there is electricity, there isn't electricity) the binary calculation, namely what is written in the software.

Hence, higher software language, which is nothing but a procedure that defines certain actions done on physical symbols, is subject to *entire and precise* breakdown into elementary and simple logical-mathematical routines based on the binary calculation (0 and 1), which the computer performs physically, electrically. Every higher computational procedure (algorithm), then, is defined wholly by sub-routines, which in the end are performed actually and completely by physical actions. By means of these mechanical procedures it is possible to develop different programs, which, as I mentioned above, are capable of doing calculations of different kinds, of adding up and multiplying numbers, of solving mathematics equations, of word-processing, and of drawing static and dynamic pictures.

Now, let us return to our main question concerning the analogy to the computer: is the consciousness-brain connection analogous to the software-hardware connection on which the computer rests? A number of researchers and philosophers believe that the answer is affirmative (see, e.g., discussions and critiques in Block, 1995; Cummins, 1983; Dennett, 1979; Rakover, 1990). The idea underpinning this answer is the notion of breaking down an intricate system into its parts: just as software can be disassembled into simple components (simple routines), which may be disassembled into still simpler parts, until the simplest, the most primitive, units are obtained, such as the units of *and*, *or*, and *not*, binary units that decide between 0 and 1, two states that may be defined by means of two electrical states, so is it possible to decompose every cognitive process, cognitive property, to the most primitive level that may be connected to the neurophysiology of the brain. And just as the computer actually runs the program by means of its electrical action, so the brain runs the cognitive system by its neurophysiological action. Dennett (1979), describes this decomposition idea as:

> … a *top-down* strategy that begins with a more abstract decomposition of the highest levels of psychological organization and hopes to analyze these into more and more detailed smaller systems or processes until finally one arrives at elements familiar to the biologists. (p. 110)

Dennett goes on to describe this procedure of decomposition in an entertaining way. He describes a mental process by means of a flow chart made up of small boxes interconnected by arrows, where every box renders a "homunculus" with certain functions. He continues:

> If we then look closer at the individual boxes we see that the function of each is accomplished by subdividing it via another flow chart into smaller, more stupid homunculi. Eventually this nesting of boxes within boxes lands you with homunculi so stupid (all they have to do is remember whether to say yes or no when asked) that they can be, as one says, "replaced by machine". (p. 124)

Does this description indeed hold? To my mind, the process of decomposition of behavior carrying meaning, conscious behavior, is very difficult, and it is not easy to see, in the end, how such behavior may be connected to simple neurophysiological states. Several barriers block the application of the decomposition process.

– While decomposition is based on a mechanical system such as the computer, the consciousness/brain system is not perceived as a purely mechanical system. For example, in the mechanical system it is possible to point to its basic building blocks, but it is not easy to do so in the consciousness/brain system. Therefore, application of a mechanistic decomposition process to a system that is not wholly mechanistic may run into problems whose solution is not in sight. Fetzer (2001) in his book *Computers and cognition: Why minds are not machines* says something very similar: consciousness is not a computer program because for the computer the symbols of the software are nothing but meaningless physical symbols, and because consciousness works on conscious symbols in a manner entirely different from the algorithmic manner in which the computer works on physical signs.

That is, a mechanistic process of decomposition applied to the consciousness/brain system is liable not to be suitable because of these differences.

- Goldstein (2005) reviews in her book on Gödel's incompleteness theorems some interesting implications of these theorems about the question whether the mind is a computer. The first theorem proves that in an axiomatic system, which includes arithmetic, there are certain true arguments that are not provable in this system. The second theorem proves that it is impossible to prove the consistency of such a system. Several researchers (such as Lucas and Penrose) argued that if a basic mathematical system is not given to complete logical comprehension, then clearly the mind cannot be understood as a formal system like the computer. That is, one cannot reduce the mind to a formal system of mathematical rules.
- It is hard to see how the conscious emotional-thought experience "I love Max the cat" can be broken down into simpler modular parts, to the basic binary computational level, which may be linked to neurophysiological states in the brain, without losing information, without losing the whole conscious meaning expressed in this sentence. For example, even breaking this sentence into separate words will cause every word immediately to lose the meaning of the sentence as a whole. That is, the meaning of each word separately differs from its meaning in the context of the whole sentence. The word 'love', for example, has different meanings in different contexts: love Aviva, love the state of Israel, love eating an egg sandwich, etc. By contrast, the decomposition process in the computer loses no information, and the translation from language to language down to the language of the machine is complete (otherwise the computer would not be a computer). It is not possible, then, to break down a meaningful experience without warping it. Therefore, mental behavior is plainly not of the kind of phenomena that may be decomposed similarly to systems such as the flashlight or the computer. (But see a different approach in Fodor, 1976, 1980.)
- This difficulty is not characteristic only of the above sentence: it is characteristic of cognitive psychology as a whole, because the basic concept of this psychology, the concept of information, is wide open and not defined. This concept is defined fully in the computer sciences (by binary computation), but in psychology, as stated, it is unlimited, undefined, and applies to everything imaginable in the context of consciousness and of the cognitive system: syllables, words, sentences, content, metaphors, sounds, processes of judgment and deduction, etc. (see Palmer & Kimche, 1986). This system of psychological information, then, is not a closed mechanistic system, and as a result information at one stage is not preserved in the transfer to processing in the next stage.
- Even if we accept the identity of neurophysiological state and mental state, we still will not be able to explain the special properties of mental states by means of neurophysiological properties: what Levine (1983) termed an "explanatory gap" will remain. It is hard to see how one may explain the qualitative properties, say, of the conscious experience of seeing the color red by means of physical knowledge of

light and neurophysiological knowledge of the visual system (e.g., Jackson, 1982, 1986). Jackson suggested that a scientist who knows everything there is to know about seeing colors but who has spent her entire life in black and white, will finally learn to understand experientially the meaning of the color red the moment she is exposed to a colorful surrounding. That is, even if we succeed in breaking down the vision of a red tomato into its components, and reducing them at the end of this decomposition to a number of basic neurophysiological mechanisms, we shall not be able to close the explanatory gap. Therefore I believe that in this process of decomposition something important gets lost – the conscious experience of the red color of the tomato.
- In an article that summarizes and discusses the relevant literature, Looren de Jong (2003) notes several snags in the operation of the decomposition process, for example, the possibility that higher processes possess properties of a totally new phenomenon, "emergent properties", the components of which are very hard to discern; that higher processes are continuous and dynamic, so that it is not easy to see how they may be decomposed into modular parts (and on emergent properties see chapter 9).

8.4 Reduction

Another way to solve the problem of the mind/body connection is through use of scientific methodology for inter-theory reduction, that is, through an attempt to reduce psychological theory to neurophysiological theory, something known in the professional literature as psycho-neural reduction. In other words, the question is whether it is possible to explain a psychological theory (folk psychology, cognitive psychology) by means of neurophysiological theory (see, e.g., Bickle, 1998; Kim, 1998; Silberstein, 2002).

To clarify this question I shall first describe briefly and schematically the classic methodology for inter-theory reduction. One theory, which we shall call the reduced theory (T_R), is reduced to a second, more basic theory, which is called the basic reducing theory (T_B), when T_R may be derived from T_B together with certain bridging laws, where these bridging laws connect the concepts of the two theories. Usually the bridging laws are perceived as identities, for example, in the case of reducing thermodynamics to statistical mechanics, according to which it was proposed that temperature equals the average kinetic energy. In this case statistical mechanics also offers an ontological (material) explanation for the macro concept of temperature through the micro concept of kinetic energy.

In the case of psycho-neural reduction, identity between a mental state and a neurophysiological state offers an ontological explanation of mental states – these are nothing but neurophysiological states. If the bridging laws are not identical, but, for example, correlations between (variables) concepts of the two theories, then the reduction will be flimsy, because these correlations (as I noted in chapter 7) in themselves require explanations and because it is not at all clear if they are nothing but a kind of accidental generalizations. (See different approaches to reduction and discussion on bridging laws in Causey, 1972; Fodor, 1974; Nagel, 1961; Rakover,1990; Schaffner, 1967, 1993; Sklar, 1967.)

Several arguments have been put forward against psycho-neural reduction (see discussion in Barendregt & van Rappard, 2004; McCauley & Bechtel, 2001). Most of them emphasize that psycho-neural bridging laws are not possible. For example, Davidson's (1980) argument, called 'mental anomalism', which was described above, shows that such a bridging law (psycho-neurological law) is not possible.

One of the most forceful arguments against psycho-neural reduction is that of multiple realization, noted above (see Fodor, 1874, 1998. And see a different view and debate in Bechtel & Mundale, 1999; Bickle, 1998; Looren de Jong, 2003). To substantiate this argument, we shall examine the behavioral state of pain. According to functionalism, pain is a mental property that can be realized by material processes such as various neurophysiological processes found in a large number of organisms (humans, dogs, cats, fish, reptiles, etc.), and also the electrical processes located in highly advanced and elaborate computers. On the assumption that this argument is correct, it becomes impossible to reduce a psychological law, theory, to a neurophysiological law, theory, because it is not possible to find a bridging law that will join through identity the state of pain and a certain neurophysiological state (which the theory of identity, described above, aspires to do). If this is the case, it is not possible to find a bridging law by means of which it would be possible to reduce psychological theory to neurophysiological theory.

As a counter-argument, supporters of the reductionist approach proposed the disjunctive bridging law: pain is a disjunctive realization of material (neurophysiological) states: NS_1 or NS_2 or … or NS_i. But a disjunctive bridging law, quite simply, is not deemed a scientific law. For example, which NS is suitable for the realization of pain? Are all material (neurophysiological, etc.) processes in the world suitable for realizing pain? What then are the properties of realization of NS?

A further argument against psycho-neural reduction is based on the requirement of equality of units, which I described above (see Rakover, 2002): it is not possible to discover a psycho-neural bridging law because this law does not fulfill the requirement of equality of units. According to this requirement, the units of measurement on the two sides of the law's equation must be identical. However, the measurement units of the psychological concepts are entirely different from the neurophysiological measurement units, and no common measuring standard can be found for them that will unite the psychological with the neurophysiological. While the concepts of the neurophysiological theory are measured by means of electrical, chemical and molecular changes, cognitive theories are expressed in actions measured chiefly by frequency of correct responses and speed of response. So it is hard to see how a bridging law may be built between the concepts of these two theories. The combination of measuring units of the chemical changes is not equal, for example, to the measuring unit of the response-speed index, mainly because this index expresses psychological, not physical, time. Also, response speed is an index that expresses actions of a large number of different cognitive processes (connected to perception, remembering, and deciding) which work linearly and in parallel (see discussion in Pachella, 1974). Finally I should note that if by psycho-neural reduction one intends principally to reduce the theory of folk psychology to neurophysiological theory, according to the methodological dualism approach this position is not possible because, as I showed in chapter 7, folk psychology cannot be taken as parallel to a scientific theory: it is a collection of schemes for giving explanations in everyday life. (The fact that explanations of this kind have been given throughout human history does not attest that this theory is bad, that it does not develop with scientific research, but that refutations of specific teleological predictions (desire/belief) have no empirical implications that clash with the explanation schemes themselves, which continue, therefore, to operate over many generations.)

Akin to the attempt to revive type identity theory, which I outlined above, in recent years an effort has been made to revitalize the possibility of performing psycho-neural reduction (see especially Bickle, 1998). One of the fundamental notions of this approach, which Bickle calls the "new wave", attempts to bypass the need to use bridging laws through the application of a methodology of theoretical models, which leads to the construction of a theory that is analogous, similar, to the reduced theory T_R that is marked T^*_R. According to the classic approach, one derives from the basic reducing theory T_B the reduced theory T_R by means of bridging laws; but according to the new wave approach, it is possible to create on the basis T_B a theory similar to the reduced theory T^*_R without using bridging laws. The question, of course, is how similar are T_R and T^*_R. This question led to further theoretical developments, which I shall not deal with here. In sum, the new wave approach gave rise to a great amount of criticism, which questioned the idea of psycho-neural reduction (see summary and discussion on these subjects in Barendregt & van Rappard, 2004; McCauley & Bechtel, 2001). For example, Bontly (2000) finds that even the new wave cannot escape ontological bridging laws, formulated one way or another:

> Even the 'new wave' reductionist must first determine whether mental properties can be identified with or built up from neurobiological parts, and as far as I can see, that's just the traditional mind-body problem all over again. (p. 904)

McCauley and Bechtel (2001) argue that the new wave approach does not take into account the fact that scientific development is accomplished by research cooperation between two theories on different levels of description and explanation, for example, cooperation between cognitive theory and neurophysiological theory. They propose a strategy of "hypothetical identities" as hypotheses that connect the concepts of theories of different levels whose chief function is to stimulate and encourage research, for example, between cognitive theory for processing visual information and local neurophysiological processes in the brain (V4 area). Barendregt & van Rappard (2004), who rely among others on McCauley and Bechtel, stress that inter-theory reduction is not a theory about the mind/body relationship but a methodological position in science that seeks to create a bridge between theories on different levels, and to pave the way for conducting scientific research.

8.5 Multiple realizability and decomposition – a methodological note

The main argument I wish to raise here is that these two approaches, multiple realizability and decomposition, are not in harmony methodologically. To show this I shall first examine what is the goal of science by seeking out the desirable relation between explanation and observation from the scientific viewpoint, that is, I shall look for the relation that encourages the advance of scientific knowledge and offers an understanding of development of science as we know it today; then I shall support my argument in light of this discernment. Finally I shall try to resolve the disagreement. Four possible relations exist:

(1) One explanation – one observation: Is the purpose of science to provide one sole and special explanation for every observation? I believe that it is not, because the number of different explanations will be equal to the number of observations. In this case no motivation exists to test the effectiveness of the explanations because the moment some explanation is given for some observation, scientific work is over. Furthermore, it is not clear how we may decide among different explanations for similar observations, or among different explanations that are liable mutually to contradict.

(2) Several explanations – one observation: Is the purpose of science to provide different explanations for every observation? I believe that it is not. The situation in this case is even worse than in the foregoing situation (1), because such science is liable to contain a huge number of explanations that will probably contradict, even in respect of the same observation. This multiplicity may well be grasped as an index of originality, but this originality is sterile and does not lead to anything but inner conflict.

(3) One explanation – several observations: Is the purpose of science to provide one explanation for different observations? I believe that it is. In this case science offers a uniform and general explanation for a large number of observations. This situation allows and encourages an examination of a uniform explanation by its application to additional observations, an examination that allows the change or replacement of the explanation by a more efficient and better explanation. In fact, this empirical examination is that which allows a test of whether a scientific law is nothing but an accidental empirical generalization. This methodology is what ensures the advance of research toward the great scientific ideal: to find a uniform theory that will explain all possible observations.

(4) Several explanations – several observations: Is the purpose of science to provide different explanations for different observations? I believe it is not, because such a

methodological ideal is in fact liable in the end to draw us back to possibility (2): several explanations – one observation. That is, this situation harbors the danger that science will consist of a collection of different explanations for different observations, when a number of different explanations also apply to the same observation. For example, one research program proposes that explanations *a, b, c* apply to observations 1,2,3,4, and another research program proposes that explanations *d,e,f,g* apply to observations 2,4,5,6. Hence, explanations *a* to *g* apply to the same observations 2 and 4.

The answer may be affirmative on condition that we see the present possibility as an intermediate situation meant to serve possibility (3): one explanation – several observations. From the theoretical and empirical aspect, it is very hard to reach an all-explanatory theory. It is easier and more practicable to try to develop theory A, which explains many observations in one area, theory B which explains observations in another area, and so on; then to try to unify these theories from different areas under a supreme roof theory. This, in fact, is one of the important routes that modern science is taking, for example, Newtonian theory, which explains observations and laws from different areas, such as Galileo's law of free fall and Kepler's laws of the movement of the stars. (But see other approaches too, such as that of Cartwright (1999), which in my opinion is more suited to possibility (4): several explanations – several observations. But note that Cartwright's analysis may be conceived of as more descriptive (an analysis that describes the existing research situation) than normative (an analysis that describes what ought to be done, the goal of science).)

In light of this analysis, I now move on to support the argument I noted above. I believe that multiple realizability is analogous to possibility (2): several explanations – one observation. This is because one observation of one mental state receives several explanations, that is, a large number of material factors that realize this state. By contrast, decomposition suits possibility (3): one explanation – several observations. This is because we anticipate that the decomposition of different behaviors will lead us, ultimately, to several primitive, basic, factors, by whose means it will be possible to assemble the range of all the studied behaviors. (Incidentally, token identity theory suits possibility (1): one explanation – one observation, and according to the present discussion it seems that this theory too does not meet the criterion of the goal of science.) On the assumption that this analysis indeed holds, it transpires that while the process of decomposition follows the goal of science, possibility (3), the process of multiple realizability, does not. How may this disagreement be resolved?

I think that the fundamental difference between decomposition and multiple realizability lies in the difference between what is desirable from the scientific viewpoint and what exists. Decomposition is a recommended methodology on how to do science, how to build a theory and find the elementary factors that explain given phenomena; but multiple realizability essentially describes the observational phenomena according to which different aims are achieved in different ways, instruments are constructed by different means in order to realize different aims, and the same mental state appears in different creatures possessing different neurophysiological structures.

In other words, while decomposition is nothing but a methodological guideline that accords with possibility (3): one explanation – several observations, multiple realizability is nothing but an interesting observational phenomenon which requires an explanation; it is not some sort of methodological instruction that this is how science should be done (and for multiple realizability being an empirical generalization see Fodor, 1988; Kim, 1993).

If this analysis holds, it is possible to apply the decomposition method to the phenomenon of multiple realizability, with the aim of abstracting those basic neurophysiological components common to different creatures existing in the same mental state. And if indeed this research description is possible, the spearhead of multiple realizability as an argument against identity theory, functionalism and reduction is greatly blunted.

8.6 Consciousness

As may be seen from the four subjects briefly surveyed above, not a single one of them is not interwoven one way or another with consciousness. This is simply because consciousness is an essential part of the mind/body relation. These four subjects are not all the areas of research in consciousness; however the discussion of all these areas is beyond the purposes of the book (see reviews and discussions in Chalmers, 1996, 2003; Dennet, 1991; Siewert, 1998; Tye, 1996). A large and principal part of research on consciousness centers on the question of how may consciousness, with all its properties and kinds, be explained (see, e.g., Van Gulick, 1995); but in the present section I would like to concentrate not on the explanation of consciousness but mainly on the argument that consciousness is a vital factor in understanding human and animal behavior. The justification for this argument stands on two legs: the empirical leg, which presents us with a collection of phenomena of behavior that are difficult to explain without employing consciousness and mentalistic explanations; and the philosophical leg, which proposes that it is hard to treat consciousness as an epiphenomenon. In earlier chapters I provided many observations (involving Max the cat) empirically supporting the necessity of a mentalistic explanation; here I shall concentrate mainly on the philosophical context, and finally I shall discuss additional empirical evidence.

In the context of the philosophical discussion, I have to admit that even though the methodological dualism proposed in the book does not require the development of any speculation on the subject (because in essence it is no more than a methodological proposal for giving mechanistic and mentalistic explanations for complex behavior), I found myself examining the philosophical whirlpool around the subject from a theoretical angle, a viewpoint that I call "consciousness-induction" – an approach that says a little more than everyday knowledge about the mind/body problem that I set forth at the start of this chapter. The hub of this approach is a "consciousness-inducing process", based on the theory of Schacter (1989): "Dissociable Interactions and Conscious Experience (DICE)", and on McGinn's (1991) suggestion that "... the brain has *some* property which confers consciousness upon it; I do not say that I know *which*" (pp. 204-205). First I shall very briefly present DICE and then the consciousness-induction approach.

According to DICE, various modular units of memory enter into interaction with a conscious awareness system (CAS), which determines their kind of conscious experience. In the cases in which memory information does not succeed in reaching the CAS, it can still influence the individual's responses without awareness. Furthermore, on the basis of neuropsychological studies Schacter suggests that the posterior parietal cortex in the brain is involved in conscious experience.

Following the above authors, I suggest that a consciousness-induction process works in the neurophysiological-cognitive system, whose function is precisely what its name indicates: to induce conscious experience of different kinds of cognitive representations, cognitive information, which are realized by neurophysiological activity in the brain. (Although the concept of information in psychology is wide open and undefined (see discussion in chapter 4, and in Palmer & Kimche, 1986), I cannot suggest an exact characterization of this concept, so I shall continue to use it in the accepted senses in psychology and the cognitive sciences.) The kind and degree of conscious experience are determined by the kind of cognitive representation that enters into interaction with the process of consciousness induction. For example, the conscious experience of pain is different from that of tickling because the representation of pain differs from that of tickling; likewise the conscious experience of seeing differs from that of hearing because these sensory systems create different representations; and the conscious experience of the sentence 'I love X' differs from the conscious experience of the sentence 'I hate X' because there is a linguistic-content distinction between these two verbs. (It is worth noting that while according to my approach the kind of conscious-experiential content is determined by the senses and the different representations, DICE theory posits that the kind of conscious experience is determined by the CAS itself: "… the conscious system defines a particular kind of conscious experience" (p. 363).)

A proposal of this kind does not answer a good number of important questions (e.g., what is the nature of the consciousness-inducing process, and how does it work? What are the properties of the events, processes that allow induction of consciousness?); nevertheless, this proposal draws explanatory outlines

- for separation of consciousness and intentionality, which I described in chapter 5, that is, for observations where in most cases (e.g., daily life) intentionality exists consciously even though in some cases intentionality exists non-consciously;
- for phenomena of implicit learning and memory, which are different from learning and memory in a state of awareness;
- for supporting the argument that the initiation of a cognitive-mental-conscious process takes place on a non-conscious level (see discussion in Libet, 2002; Rakover, 1996; Velmans, 1991);
- for changes in behavior as a result of the presence – absence of information in consciousness. For example, everyday phenomena whereby certain information

> has flown from the memory, from consciousness, and therefore we do not perform the action that is required by it: David does not show up for the meeting as expected, because the information "Meeting fixed for one o'clock" has slipped out of his consciousness. And another example: Breznitz (1989) showed that the very giving of precise information on the duration of performance of a difficult task (a strenuous and speedy journey, continuous pressing on a dynamometer, inserting a hand into ice-water) has a dramatic effect on the level of performance of the task. Subjects in the "with-information" group did better at performing the task than subjects in the "without-information" group: the number of subjects who completed the task in the with-information group was far higher than the number who did so in the without-information group. Findings of this kind support the interpretation that consciousness has an important function in the explanation of behavior. If indeed conscious experience was an epiphenomenon, it would be hard to understand such findings.

Finally, it is worth noting that the consciousness-induction approach has several interesting implications for the following subjects.

(A) Hard and easy problems of consciousness: Chalmers (1996, 1997a) suggested distinguishing two kinds of consciousness problems: easy problems and hard problems. He defines these problems thus:

> The easy problems of consciousness are those that seem directly susceptible to the standard methods of cognitive science, whereby a phenomenon is explained in terms of computational or neural mechanisms. The hard problems are those that seem to resist those methods. (1997a, p. 9)

The hard problem is expressed in the enormous difficulty in understanding how cerebral, neurophysiological, states become consciousness states; how conscious phenomena, private conscious experiences, are reduced or stem from a neurophysiological system. The easy problem is connected to cognitive behavior, for example, responding to an environmental stimulus, distinguishing, organizing information, reporting on mental states, concentrating attention, and voluntarily controlling behavior. These behaviors are explicable, in Chalmers' opinion, through cognitive and neurophysiological mechanisms, that is, through mechanisms that realize the function of the studied behavior, for example, the function of response to an environmental stimulus, of discrimination, and so on.

By means of this distinction between the easy and the hard problem the failure of various proposals to solve the problem of consciousness by appeal to neurophysiological processes in the brain becomes understandable: the aim was to crack the hard problem but the result was (in the best case) a sketch for solving the easy problem. We may take as an example the proposal of Edelman & Tononi (2000). They attempted to explain the hard problem, how consciousness arises as a result of specific neural processes and how qualia are understood in terms of these processes, but in essence what they propose is outlines for solving an easy problem, how to explain neurophysiologically two basic properties of consciousness as complex cognitive processes. On the one hand, these processes are integral and cannot be divided into their components, and on the other hand they are informative and include an enormous number of different and distinct mental states. These properties are explained, in their opinion, by certain distributed neural processes that display precisely these properties of integrality and distinctiveness. That is, consciousness characterized as complex processes is realized by this kind of complex neurophysiological processes in the brain. Despite the great interest that the proposal of Edelman & Tononi arouses, the hidden exceeds the revealed: while their proposal can be seen as a fascinating attempt to explain an easy problem, the basic problem, the hard problem, is not solved. It is not clear how these neural processes bring about consciousness. The mind/body explanatory gap is still to be bridged over.

I agree with Chalmers that the hard problem is indeed hard, but I don't agree that the easy problem is indeed easy. To my mind it is no less hard than the hard problem, for the following reasons. As I repeatedly stated above, a complex behavior (e.g., learning, perceiving, concentrating attention) is the carrier of a conscious experience imparted, according to the above, by a 'consciousness-inducing process'; an experience that bestows on a behavior the unique, characteristic meaning of each and every person: a sense of reality, an experience of being, conscious existence. Without consciousness, responses, actions, behavior would be purely motor movements; we would be nothing but robots, machines, moving in the physical world, like the movement of stars, electrical discharges and the reflection or absorption of light. We would be explainable by a collection of physical, chemical, and biological laws alone; and in truth, there would be no point to living, no motive or reward for what we want, believe, and do. The explanations that Chalmers suggests for the easy problems are nothing but mechanistic explanations, as may be seen from the explanations he offers for access to, and reportability of, mental states:

> To explain access and reportability, for example, we need only specify the mechanism by which information about internal states is retrieved and made available for verbal report. (1997a, p. 10)

In his opinion, the appropriate explanation is given by a description of the (cognitive, neurophysiological) mechanism that performs the function of the studied behavior. These explanations, by their very nature, leave no room for consciousness as an explanatory factor – consciousness is an epiphenomenon. Similarly, Harnad (2000) writes:

> The functional stuff would all go through fine – behaviourally, computationally –*if* we were all just feelingless zombies. But we're not. (p. 56)

This is the state of affairs (employing computational explanations only) that this book opposes. The approach of the multi-explanation theory is that a mechanistic explanation alone cannot contend with complex behavior involving consciousness. Without using, in addition to the mechanistic explanation, a mentalistic explanation too, the mechanistic explanation will be a partial explanation only – an explanation appropriate for physical-neurophysiological robots. The mechanistic explanation can detail the physical-physiological mechanism responsible for raising the hand, but in no way can it offer an explanation for the most important thing of all, for the questions why was the hand raised? What is the meaning of raising the hand, its purpose? And the question of how it happens that a motor movement turns into an instrument for expressing a mental state, a process; that is, how does it come about that a certain motor movement becomes an instrument in the service of the mentality? I suggest, therefore, that meaningful behaviors are a kind of executive branch of the mental system. These behaviors should not be seen as a kind of motor behavior to which consciousness has been added, but the reverse: this behavior, its very essence, is the conscious meaning that has been induced in it.

Chalmers' (1997a) article is a target article, which gave rise to a large number of critical papers, and to Chalmers' (1997b) response to them (see the book edited by Shear, 1997). Some of these critical papers, which Chalmers (1997b) terms "nonreductive analyses", assert, like what I have argued above, that the easy problems are no less thorny than the hard problems. I shall not go into a discussion of these matters, but move on to illustrate, by means of analysis of one more behavioral episode of Max the cat, namely **multi-response learning**, that the easy problem is not that easy, that is, the mechanistic explanation may well constitute in the best case a partial explanation for a complex behavior.

Max has a delicate stomach and from time to time suffers attacks of diarrhea. When this happens we lock the cat in the kitchen at night for about two weeks (until we are certain that the medical treatment has worked and the diarrhea has completely disappeared). Aviva usually goes to bed before me, and I, who am late to bed and late to rise, before retiring generally first close the bathroom door leading to the porch in common with the kitchen, return after fifteen to twenty seconds to the living room, go to Max, lying dozing on Aviva's armchair, carry him, while stroking him, to the kitchen, and close the door on him. After two or three times Max learned that the closing of the bathroom door meant his being locked in the kitchen for the night. This learning was expressed in three categories of response, whose frequency at a very rough estimate was (a) 60%, (b) 25%, (c) 15%:

(a) *Hiding*: Max jumps down from the armchair and tries to hide in places where it is hard to grab him, mostly under the table on the porch.

(b) *Tenseness*: Max stays on Aviva's armchair in a state of great tension – his body is tense and his ears are pricked up.

(c) *Going to the kitchen*: Max jumps off the armchair and goes to the kitchen, or waits a short while at the threshold of the kitchen. (I never found him going to the water or food placed at the end of the kitchen.)

How may these behaviors be explained? My argument is that it is very hard to suggest a mechanistic mechanism that will explain these three kinds of response. In my opinion, this behavior meets the criterion of mentalistic behavior, free will, which I developed in chapter 3: ***the same individual, whose private behavior has changed***, in the same state of stimulus, at different times, responds with ***different responses***. As is evident from the present case, even though the state of stimulus did not change, Max responded at different times with entirely different responses. As a result, I suggest that what changed in Max is his mental state, that is, the cat's free will. To facilitate understanding the answer to the question: what made Max try to hide one time, to stay where he was, tensed up, another time, and "to accept his fate" and go into the kitchen yet another time, we shall look at a similar episode, "doing homework", which happened in my childhood when I went to elementary school.

At that time my mother made the rule of homework: first you do your homework and then you can play. My responses to this situation were split into three categories: (1) follow mother's rule; (2) sneak out of the house to play and slip back in toward evening to do my homework; (3) pretend to read a schoolbook while actually reading adventure stories or thrillers (at that time there was still no television in Israel). Can this episode be explained mechanistically, by means of some cognitive-neurophysiological mechanism? I don't think so. Personally I have no doubt that I was entirely responsible for these responses of mine – out of my free will to obey or disobey mother's homework rule. I believe that a similar explanation may be suggested for **multi-response learning**: Max learned quickly the "sleep-in-the-kitchen rule": you are going to sleep in the kitchen the moment the bathroom door closes. If he had not learned this rule it would be hard to explain even one of his three responses: hiding, tenseness, and going to the kitchen. Hence the going-to-the-kitchen response may be seen as following the sleep-in-the-kitchen rule, while the other two responses constitute infringement of this rule.

The point I want to emphasize is that it is hard to stitch onto Max's behavior a mechanistic explanation. As we are dealing with **multi-response learning** here, let us look for a simple learning explanation. It is reasonable to suppose that locking the cat in the kitchen is a negative reinforcement, a punishment (in the morning, when we open the kitchen or the bathroom door, Max gets out of there fast) so the cat's attempt to hide may be regarded as an avoidance response. This learning may explain 60% of Max's behavior! But what about the two other responses? If these responses could be categorized in the framework of the concept "avoidance response" they might possibly be regarded as part of the chain of responses that eventually lead to avoidance (for example, the response of approaching the lever in avoidance learning in the Skinner box). It is thus possible to see (with much good will) Max's tenseness response as a kind of freeze response characteristic of learning fear and avoidance (see, e.g., Rakover, 1975,

1979, 1980), although my impression is that Max's tenseness response is not of this kind (e.g., the cat did not crouch, did not emit an uncontrolled yowling, his eyes were not wide open, his hair did not stand on end, and his ears were not flattened back). But how may we explain the response of going to the kitchen in the framework of learning avoidance? To be frank, when these responses appeared I was greatly surprised.

Furthermore, on some nights I put Max in the kitchen, closed the kitchen door, went through the lounge into the bathroom, and closed the door. Max could have taken the opportunity of my time going to the bathroom to escape; however, he did not slip out of the kitchen through the bathroom, but sat under the kitchen table, looking at the open door, until I came and closed the bathroom door. It is thus reasonable to suppose that in this case Max submitted to the sleep-in-the-kitchen rule.

(B) Consciousness in animals: I assume that a consciousness-induction process exists in animals also, where the degrees and kinds of consciousness induced depend on the cognitive representations that are created by the appropriate neurophysiological systems.

This approach comes up against the following question, which I partly discussed before: how may one know that animals have consciousness similar to ours? This is a question connected to the "other mind" problem: if only the person herself has access to her consciousness, how may she know the consciousness of the other? How may she know that the other has awareness, consciousness? Since I discussed this problem in the earlier chapters, here I shall say only the following things.

The answer to this question may be formulated according to two kinds of game-rules. According to the game-rules of *logical proof*, my impression is that the question has no answer. I have not found anywhere a proof in principle that other people and animals (such as Max the cat) have or do not have consciousness. Most of the reasoning for or against consciousness in animals is based on analogies, for example, the following: A is endowed with qualities *a, b, c, d, e*, and therefore if B is endowed with qualities *a, b, c, d* we infer that B is also endowed with quality *e*. However, analogies are a kind of inductive inferences, so we should not be surprised if B is not endowed with quality *e* even though he is endowed with qualities *a-d*. For example, A is a person endowed with neurophysiological-behavioral qualities *a-d*, and also with quality *e*, which is conscious; B is a zombie endowed with qualities *a-d* but without any trace of consciousness. (Thus, as I argued above, I believe that the zombie, the robocat, arguments are based on analogies.)

Analogies are part of the basis of the debate over whether animals are endowed, like us, with self-awareness. Gallup (1998), who invented the mirror test for self-recognition in a mirror, maintains that chimps, like us, are endowed with self-awareness, with self-consciousness, because they pass the mirror test (like us, chimps look into a mirror and are able to discern that some change has occurred in their face); but Povinelli (1998) does not think so, and criticizes Gallup's experiments. Is Max likewise endowed with self-awareness? I do not have observations supporting this hypothesis: I have not discovered evidence in Max's everyday behavior supporting self-recognition, even though Max, like all cats, licks his body endlessly, and occasional also peeks into the big mirror in the lounge. To my mind, his behaviors support mainly mentalistic explanations based on conscious intentions.

According to the rules of the *scientific game* the story is entirely different. According to these rules, science does not supply ultimate proof of the correctness of the hypothesis, but offers it deep empirical-theoretical grounding. We hold theory (a) and not theory (b) because the first theory gained greater theoretical-empirical grounding that the second. But theory (a) too is liable to be rejected and replaced by theory (c), when the degree of grounding of theory (c) exceeds that of theory (a). In this respect, I suggest, the hypothesis that David has conscious-mental states similar to mine is an extremely useful specific hypothesis, because so far David's behavior matches what is implied by it (e.g., David behaves according to the desire/belief explanation scheme). In the same way, I may add that the specific mentalistic hypotheses that I gave in earlier chapters for understanding Max's behavior are extremely useful because they were well supported by the cat's behavior.

Moreover, it may be suggested that speculation about consciousness induction takes the question of the "other mind" close to the region of the scientific discussion, because it is reasonable to suggest that the consciousness-induction process is an evolutionary outcome, hence transfers the question of consciousness in animals to the question of the neurophysiological mechanism involved in imparting of consciousness (see a similar idea in Dawkins, 1995).

(C) Why is the concept consciousness necessary? As I argued above, we need the concept of consciousness to explain the effect of the presence–absence of conscious information on behavior. By contrast, the epiphenomenal approach, which I discussed earlier, suggests that consciousness has no effect on behavior. Because of the importance of the subject, I discuss this matter here too. Like the idea of conscious inessentialism (see Flanagan, 1992, above) Dawkins (1995) writes:

> Even in principle, however, no comparable experiments appear to be possible for detecting the presence of conscious awareness. There is no prediction we can make that if the animal has consciousness it should do X but if it is not conscious it should do Y. (p. 139)

If this argument is valid, then the methodological dualism approach is bankrupt and all that remains of it is a certain practical value that suggests that the mentalistic explanation should be used until a better mechanistic explanation is found. Assuming that the previous chapters supported at least the practical value of methodological dualism, we shall now examine Dawkins' argument from a theoretical and empirical viewpoint. (Dawkins tries to give an answer to our question by finding adaptive functions fulfilled by consciousness.)

- Let us move Dawkins' argument to the domain of robotics, and realize it by constructing a robocat, a physical double of Max the cat, that behaves exactly like

Max, but is devoid of consciousness. According to Dawkins, it transpires that the fact that Max is endowed with consciousness, and the robocat is not, makes no difference – it has no expression in behavior. On the one hand, this argument is circular, because if it is assumed from the outset that the robocat's behavior is identical to that of Max, who possesses consciousness. The argument assumes that consciousness induction to the computer-brain of the robocat, or the removal of consciousness from Max's brain, will make no difference, because precisely the absence of this influence of consciousness was assumed already in the argument's realization – in the very creation of the robocat. On the other hand, if we do not assume such a close identity as this between behavior with consciousness and behavior devoid of consciousness, then a difference in behavior will remain, hence a place for the influence of consciousness.

- Let us construct such a robocat and look, for example, at the process of recognition, identification, of a dog by the robocat. Into the robocat's computer-memory let us introduce a collection of properties of a dog (four legs, a body, a head, hair, size, barking, etc.) which will be arranged in the form of a vector that we shall designate V_m (memory vector); now let us assume that the robocat encounters a dog whose image is translated into a perception vector, V_p. The robocat's computing system compares these two vectors and finds them identical. Now the question arises, what will the robocat do when its computing system finds $V_m = V_p$? In my view the answer lies with the programmer, who decides what the robocat's appropriate response will be! And this response will be *nothing but a kind of interpretation by the programmer of the situation cat meets dog. That is, the robocat's response will be nothing but a kind of realization of the programmer's conscious meaning!* (It does not matter if the appropriate response, to run away, is determined in advance, or if the robocat learns through changes in the neural network and the results of its computations are linked to the appropriate programmed movements.) This interpretation that I suggest accords with the approach of Leibowitz (1982) who argued that only in respect of the person possessing awareness and consciousness is there meaningful content to the physical symbols printed out by the computer.
- The last point takes Dawkins' argument to the region of the discussion of empirical methodology, which holds that for every phenomenon a large number of alternative explanations may be proposed, which must be put to the test of reality. And the question that arises now is, has empirical evidence indeed been found showing that without the mentalistic explanation it would be hard for us to suggest an efficient explanation for the individual's behavior? The answer, I think, is affirmative, and the discussion of Max's behavior, for example, in the previous chapters will attest to this. Moreover, as we saw above, there is a variety of empirical evidence showing that it is possible to change the degree and kind of consciousness (e.g., as a result of brain damage, local anesthesia, drugs, hypnosis, amnesia, implicit memory) that causes behavior different from the behavior of a conscious being in a normal state.

CHAPTER 9

Methodological dualism and multi-explanation theory in the broad philosophical context

In this chapter I shall try to elucidate the unique nature of the approach I have developed here by comparing it with relevant approaches in the philosophy of science and the mind. First I shall compare methodological dualism (and the Scientification method: giving scientific methodological sanction to mentalistic explanations) with explanatory dualism, functionalism, and levels of explanation; then I shall compare the approach of multi-explanation theory with other approaches to constructing a theory; finally I shall discuss the following questions: what kind of understanding does this theory supply? What is its relation to consciousness viewed as emergent property?

And after all these years that I have devoted to reading books, to plowing through articles, and after all this terrible effort, which ground my brain/mind exceedingly fine, and which finally yielded a new methodology for constructing theories for understanding the behavior of animals, Max, this beautiful cat, walked into my room, let out a long yowl and pissed on all the books, articles, and notes spread across my desk. Go figure.

9.1 Methodological dualism, Scientification, explanatory dualism, functionalism, and levels of explanation

As I argued in the last chapter, methodological dualism differs from substance dualism, which assumes that the mind is a different substance from the body, and also from property dualism, which assumes that the mind and the body are two different properties of the same substance. Methodological dualism does not make ontological assumptions of that kind, but, as its name indicates, it tries to suggest a scientific methodology that unites an explanation of mentalistic, conscious behavior with an explanation of mechanistic, neurophysiological, motor, and cognitive behavior. In this respect the present approach is close to explanatory dualism, which Sayre-McCord (1989) characterizes as an explanatory difference:

> The explanation of nature is fundamentally different from the explanation of action. (p. 137)

A considerable portion of studies on explanatory dualism attempt to establish this explanatory difference, an effort that I referred to in the earlier chapters: compared with the mechanistic explanation, the mentalistic explanation is anchored to consciousness, to social norms, and to practical, rational, functional inferences, and it displays a conceptual connection between the explanans and the explanandum (see discussion in Brook & Stainton, 2001; Maxwell, 2000; Sayre-McCord, 1989).

Here I shall discuss the approach of Maxwell (2000), who supports explanatory dualism. According to him, the scientific explanation is not able to deal with conscious experience just as a blind man cannot grasp the essence of the experience of seeing the color red. (Incidentally, Maxwell argues in this article that he wrote articles preceding the works of Jackson, 1982, 1986, and of Nagel, 1974, which addressed the present argument holding that conscious experience lies outside the sphere of discussion of the natural sciences.) To treat phenomena of this kind, use must be made of what he terms a "personalistic explanation": a valid explanation, well-founded intellectually, and not given to reduction to a scientific explanation. This is how he describes the nature of this explanation:

> Personalistic explanations seek to depict the phenomenon to be explained as *something that one might oneself have experienced, done, thought, felt.* (p. 57)

This approach, Maxwell writes, contains an anthropomorphist component, and is similar to the explanations of folk psychology, which I discussed above (see on simulation theory). The scientific and personalistic explanations are largely in conflict (when the former is understood the latter becomes not understood, and vice versa), so the attempt to construct a super-theory uniting these two kinds is doomed to failure.

Although methodological dualism, as I argued above, is close in spirit to explanatory dualism, it is far from Maxwell's approach on several points connected to the approach that I call Scientification. According to this approach it is possible scientifically to propose mentalistic explanations after these have been sanctioned, approved, and amended so that they may meet the methodological requirements of science (or an important part of these requirements), that is, after these have been brought under the overall umbrella of the science game-rules. This approach differs from the Naturalism approach, which suggests that mental phenomena should be investigated by a methodology that prevails in the natural sciences, and explained by processes of nature that are not semantic or intentional (e.g., Brook & Stainton, 2001; Polger, 2004; Von Eckhardt, 1993). While the Scientification approach looks for that change that will allow the mentalistic explanation (e.g., the teleological explanation) to be sound scientifically, the Naturalistic approach proposes that mentalistic behavior needs to be explained by approaches and processes acceptable in the natural sciences. The present approach, therefore, seeks to change procedures common in everyday explanations such that it will be possible to use them as scientifically proper explanations for mental phenomena. This mentalistic explanation, despite satisfying the methodological requirements of science, is not reducible to mechanistic explanations. In other words, the Scientification approach aims to provide scientific methodological legitimacy to explanations common in folk psychology (as explanations that were created by scientific explanation schemes, and that can be put to a scientific test), so that it will be possible to use them in the general framework of the science game-rules, and to suggest explanations for behavior founded on consciousness and intentionality, without these explanations being mechanistic in nature. Just as it is possible to suggest mechanistic explanations based on material events and processes for a huge complex of material phenomena, so is it possible to suggest mentalistic explanations founded on mental events and processes for a huge complex of behaviors that contain mental components. These two kinds of explanations, therefore, match two different kinds of behaviors; the explanations complement each other coherently in the framework of multi-explanation theory.

The Scientification approach is one of the pillars on which is built methodological dualism, which may be summarized by means of the following three statements:

a) Mentalistic explanations for behavior may be used when these are formulated as hypotheses comparable to mechanistic hypotheses.
b) Specific mentalistic explanations may be used when these are created by mentalistic explanation schemes: schemes that meet the requirements of scientific methodology that prevail in the natural and social sciences.
c) A complex behavior may be explained by multi-explanation theory which coherently unites use of mentalistic and mechanistic explanation schemes.

Methodological dualism, then, is an approach that suggests a procedure that satisfies the requirements of the natural and social sciences, and by means of which it is possible to construct a theory that seeks to propose a mechanistic and a mentalistic explanation together for the behavior of humans and animals. This approach, therefore, is different from Maxwell's in that it finds a solution to the problem of anthropomorphism (see (a) above); perceives specific mentalistic explanations not as a folk-scientific theory or as a process of psychological simulation, but as explanations based on scientifically sound explanation schemes (see (b) above); and shows that it is possible to use coherently multi-explanation theory, which offers mentalistic and mechanistic explanations for given behaviors, explanations that are not ad hoc but may be put to the test of reality (see (c) above).

Methodological dualism is also different from functionalism and its associated theoretical approaches: the information processing approach, which is widespread in psychology, the representational theory of mind, which proposes that mental states, such as desire and belief, are mental representations possessing semantic content, and the computational theory of mind, which suggests that the mind functions like a computer program that operates on physical representations according to certain syntax rules. Here I cannot go into a discussion of these intricate subjects (see partial discussion in previous chapters), and I shall concentrate on two important viewpoints: levels of explanation and realizability.

(1) According to the *level of explanation approach*, behavior may be divided into three levels of description and explanation (see discussions on this matter in Block & Alston, 1984; Fodor, 1976, 1981; Pylyshyn, 1984, Von Eckhardt, 1993):
 a) the *mentalistic level*, which refers to private behaviors such as desire, belief, aims, intention, feelings, emotions, and consciousness;
 b) the *information processing level*, which refers to cognitive phenomena and cognitive concepts anchored to analogy with the computer, for example, information processing, coding, storage, and retrieval;
 c) the *neurophysiological level*, which refers to phenomena and concepts anchored to the domain of the natural sciences, such as physiology, neurology, and biochemistry.

For example, Von Eckhardt (1993) writes:

> ... the most important levels are the folk-psychology level, the information-processing level, and the neural level. It is also usually assumed that these levels are the products of (respectively) common sense, the non-neural cognitive sciences, and the neurosciences. (p. 318)

How are these three levels related? A common answer is that the connection between them is brought about by means of the realizability relationship: mentalistic processes are realized by processes of information processing; information processing is realized by neurophysiology. But in many cases, as we saw in the last chapter, researchers speak of direct realization of mentalistic processes by neurophysiology. Fodor (1981), for example, writes:

> For if the relations in terms of which psychological kinds are functionally defined can be restricted to those in terms of which Turing machine program states are specified, the mechanical realizability of the psychological theory which posits the kinds is thereby guaranteed. (pp. 13–14)

(A Turing machine, invented by the mathematician Alan Turing, is a theoretical model that forms the basis of any computer and is capable of performing the operations of any computer.) He goes on to write: "In short, what Turing machine functionalism provides is just a *sufficient* condition for the mechanical realizability of a functional theory" (p. 14). Von Eckhardt (1993) writes:

> ... for each of the basic general properties of our ordinary cognitive capacities, with the exception of reliability, there is an information-processing property (or a set of information-processing properties) that plausibly realizes that property at the information-processing level. (p. 317)

(2) On account of the *multiple realizability argument*, which I discussed above, it may be suggested that the functional explanation, the information-processing explanation for cognitive phenomena (perception, memory), is autonomous, and does not require a description that details how this explanation actually occurs physiologically. That is,

the explanation of information processing is similar to the explanation given to the question of how the computer adds numbers by detailing the appropriate software without referring to the hardware processes (see Fodor, 1974; Putnam, 1967. See also a critique of this approach in Feest, 2003, who supports psychological explanatory autonomy alone, where this kind of explanation is based on functional analysis after Cummins, 1983).

As we shall see below, methodological dualism criticizes these points. First I shall discuss the epistemological methodological status of level (b), the information-processing level.

If the concept of realizability were uni-dimensional, it would be possible to suggest that level (b) is not necessary: if level (a) is realized by level (b) which is realized by level (c), it transpires that level (a) is realized directly by level (c), and there is no need for the intermediate level, the information-processing level. This suggestion gains support from the fact, as stated above, that several researchers speak of direct realization of the mentalistic processes by means of neurophysiology. Furthermore (as I argued in chapter seven), one may think about the possibility that some breakthrough in the natural sciences, a revolutionary discovery, is likely to replace the computational approach prevalent today, and to remove it from the area of discussion of psychology, and to suggest in its place a new conceptual-empirical framework that will advance psychology to a higher level.

Nevertheless, it does not seem that it will be possible to forgo the information processing level for the time being, for three reasons. First, speculation about a new scientific breakthrough seems premature.

Secondly, the notion of realizability is multi-dimensional, so it does not necessarily uphold the transitivity principle. In my opinion, realization of level (a) by level (b) is different from realization of level (b) by level (c). While according to the first realization the intentionality-purposive explanation for mental behavior is translated into representations carrying semantic content operated by rules of syntax, according to the second realization, as determined by token identity theory, specific cognitive states are identical to specific neurophysiological states.

And thirdly, Von Eckhardt (1993), who discusses the relation between level (b) and level (c) (the information processing level and the neurophysiological level), supports the argument that she calls the "information processing ineliminability thesis" (pp. 330-339): it is not possible to eliminate level (b) because of the argument of multiple realizability, and because level (c) is not able to explain properly cognitive phenomena, which are well explained by the information processing approach.

Even if we accept that psychology needs the conceptual framework of information processing, and level (b) should be seen as an autonomous area of discussion, it is worth raising the following points of criticism:

- According to methodological dualism, the reason there is room for level (b) is not because level (b) justifies the mentalistic level or because of the argument of multiple realizability, but because information processing is a special kind of mechanistic explanation suited to a certain kind of cognitive phenomena for which it is hard to suggest mechanistic-neurophysiological explanations.

- Similarly to the argument that procedures of information processing are autonomous because of multiple realizability, it is possible to argue that mentalistic processes are autonomous for the same reason: they are realized by different processes of information processing. That is, mentalistic explanations may be proposed for different behaviors, without addressing the question of how these mentalistic concepts and processes are realized in practice by procedures of information processing (and also by neurophysiological processes). However, according to the methodological dualism approach, the autonomous use of mentalistic explanations is justified not because they stem from their resting on cognitive, scientific, psychology, but because the mentalistic explanation has obtained scientific methodological sanction (i.e., the mentalistic explanation has undergone Scientification) as suitable for a certain kind of behaviors included on the mentalistic level. Furthermore, it may be said that in most cases cognitive psychology does not express its explanations in intentionalist language (desire/belief) but in language analogical to computer language, language of information processing. By contrast, multi-explanation theory suggests that to explain a complex behavior, which includes mentalistic components, use must be made of several kinds of explanation: mentalistic explanations formulated in mentalistic, intentionalist, conscious language; mechanistic-computational explanations, formulated in language of the analogy to the computer; and mechanistic-neurophysiological explanations, formulated in language acceptable in the biological sciences. (Is it possible to develop a super-language that will unify all these languages?)
- As stated, one of the important arguments supporting the autonomy of psychology as an independent scientific discipline is the argument of multiple realizability, according to which it emerges that it is very hard to find a systematic connection (of identity between types, of reduction) between psychology and neurophysiology (e.g., Fodor, 1974; Putnam, 1975). This autonomy may be understood from two viewpoints: metaphysical autonomy, which proposes that it is hard to reduce psychological states and processes to material states and processes, that is, it is hard to grasp mentality in neurophysiological terms; and explanatory autonomy, which proposes that psychology has an explanatory approach that differs from that of the natural sciences. Methodological dualism accepts, on the one hand, that it is hard to explain intentionality and consciousness in neurophysiological terms, but on the other hand it partially rejects explanatory autonomy, because mentalistic explanations that have undergone Scientification follow the methodological game-rules of science, even though these explanations cannot be regarded as mechanistic explanations. That is, mentalistic explanations are a special kind, because they are proper scientifically and deal with manifestations that are hard to understand in mechanistic terms. Furthermore, as methodological dualism has sanctioned mentalistic explanations as scientific without being based on

> the multiple realizability argument, this dualism bypasses the problem of epiphenomenalism that lies at the entrance to the multiple realizability argument: if the functionalist qualities of psychological concepts are realized by neurophysiological states and processes, in the end it may be said that the burden of explanation is carried by neurophysiology, so the explanatory status of psychological concepts is undermined.

Methodological dualism displays several features similar to the idea of levels of explanation (see discussions in Bechtel & Abrahamsen, 1991; Ben-Zeev, 1993; Owens, 1989). It was suggested by the levels of explanation approach described above that human behavior is amenable to description on several levels, from the neurophysiological to the phenomenological. These levels may be understood according to a number of theoretical approaches: from the complete reductionist approach, which suggests that in the end all the special sciences will be reduced to the most basic science (physics), to the completely autonomous approach, which proposes that every area has a description and an explanation suitable for it, independently of the other areas. On this matter the physicist Anderson (1972) wrote:

> The behavior of large and complex aggregates of elementary particles, it turns out, is not to be understood in terms of simple extrapolation of the properties of a few particles. Instead, at each level of complexity entirely new properties appear, and the understanding of the new behaviors requires research which I think is as fundamental in its nature as any other. (p. 393)

Methodological dualism, on the one hand, does not accept the reductionist approach, and on the other hand accepts explanatory autonomy partially, because the mentalistic explanation meets the requirements of scientific methodology, without being bound to a metaphysical approach that tries to understand mentalistic processes by means of material concepts.

Finally, I conclude the present discussion by stressing the essential difference between methodological dualism and cognitive science (see, e.g., Thagaard, 1996; Von Eckhardt, 1993). This science, which is intended to study the mind and intelligent behavior, is based on a wide range of different research disciplines: artificial intelligence, philosophy, psychology, neurophysiology, linguistics and anthropology. The principal methodological thesis of this science is this: intelligent behavior is to be understood through an appeal to structures of representations on which computational procedures operate, where these representations are analogous to knowledge structures and data in the computer, and computational procedures are analogous to algorithms in the computer. Can methodological dualism be seen as congruent in part to cognitive science? My answer is affirmative in part, because some of the mechanistic explanation schemes in multi-explanation theory are based on computation performed on representations. But other elements of this theory, linked to mentalistic explanation schemes, is not mechanistic and does not accord with cognitive explanations.

9.2 Multi-explanation theory and other approaches to constructing theories

What kind of theory is multi-explanation theory? To answer this question it is necessary to compare the structure of this theory with the structures of theory prevalent in the natural and social sciences, which I shall review very briefly. But before doing so I wish to underline the following four points connected to multi-explanation theory.

First, it is worth stressing yet again that multi-explanation theory is not a kind of specific theory on a phenomenon or a collection of phenomena, such as theories of memory or recognition of objects or faces. Nor is it a kind of philosophical theory intended to solve philosophical-empirical problems such as theories of identity or functionalism, which aspire to explain the mind/body problem. Multi-explanation theory is no more than an abstract scheme that proposes a way to explain complex behavior whose components are both biological and mentalistic.

Secondly, while the structures of scientific theories are general and abstract, and may be applied to broad research areas in the natural sciences (physics, chemistry, biology), multi-explanation theory is a structure that is proposed solely for complex behavior of creatures such as human beings and animals. The main reason for making this special structure of theory stems from the fact that the application of structures of theory prevalent in the natural sciences, all of which rest on mechanistic explanations, to complex behavior, in which mentalistic components are interwoven, does not offer an efficient and good explanation. Multi-explanation theory proposes to improve this explanatory situation by turning attention to the great importance of mentalistic explanations.

Thirdly, because, as stated, multi-explanation theory is nothing but a procedure to guide the researcher (based on the framework of methodological dualism) on how to build certain theories with the aim of describing, explaining, and testing complex behaviors, the following question arises: is this approach a purely methodological guideline, without ontological and epistemological assumptions? The answer, to my mind, is negative. I believe that purely methodological guidelines may be found in logical principles that offer us rules of inference that do not depend on some empirical area of research. The present approach is meant to address a unique kind of empirical area, so in its very nature it is bound to make several assumptions of content, for example, assumptions about mental states and processes, and about the existing and incomprehensible connection between the mind and the body, between the brain and consciousness and the mind. These assumptions, in essence, constitute the theoretical-empirical ground on which are built and applied methodological dualism and multi-explanation theory, which is proposed as the most efficient procedure for giving scientific explanations (albeit incomplete) for complex behavior.

And fourthly, the next question arises: is not a mentalistic explanation ultimately circular? As stated above, an explanation through an appeal to will/belief assumes that these states and processes are present in consciousness, so we explain conscious behavior (e.g., David travels to Tel Aviv) by conscious behavior (e.g., an intention to meet Ruth in Tel Aviv), that is, we explain consciousness by consciousness. The answer, I think, is that there is no circularity here, simply because one conscious content is not identical to another conscious content; and just as we accept that one material process explains another material phenomenon, without being jeopardized by the circularity trap, so is the situation regarding explanation of one mental behavior (that is, a complex behavior interwoven with mentalistic components), which functions methodologically as an explained phenomenon, by means of another mental process, which functions as an explanatory factor.

I now move on to review in brief three well known structures of theory that are relevant to our concern. Then I shall discuss the relation between the present approach and these structures.

The received view of theoretical structure: This is one of the most important structures of a theory proposed at the inspiration of the positivist approach to science, which in its day was the received view. This structure became subjected to profound criticism, which put forward several alternative structures for a scientific theory (see below: models and representations; see also discussion in Carver, 2002; Rakover, 1990; Suppe, 1977). The structure of the theory is divided into three main parts: a theoretical part, an observational part, and a part that joins these two. The theoretical part is based chiefly on a mathematical system (or a coherent language) and a system of concepts that give a certain interpretation, content, to the mathematical symbols. That is, the content significance of the investigated theory is not located in mathematics but is imparted to mathematics by the conceptual system, the one that invests the mathematical symbols with special content. The theoretical part is linked to the observational part by 'bridging rules' or, what is accepted in psychology, by 'operational definitions', which link part of the concepts in the theory to the stimulus state to which the individual is subject, and her responses.

Models and representations: Examination of the way scientists build theories to explain observations showed, among other things, the following two points. First, it transpired that scientists use models and analogies as means of representing certain aspects in the world with the aim of describing and explaining observations (and theoretical problems); secondly, these models may be characterized as practical, hypothetical, and idealizations (see, e.g., discussions in Giere, 2004; Rakover, 1990; Suarez, 2004; Woody, 2004; van Fraassen, 2000, 2004). A model is a system (real or abstract) that offers a certain explanatory interpretation for observations and in many cases constitutes realization of several properties in the theory. For example, the computer in psychology is proposed as a mechanistic model of cognitive processes such as perception, learning, and thinking; and a small aircraft placed in a wind-tunnel is a real and three-dimensional model that realizes a number of properties of physical theory, describing the effect of forces acting on a the aircraft's wings at different speeds. So in psychology the computer model serves as a source for explaining cognitive processes, whereas in physics the model of the aircraft in the wind-tunnel serves as a way of testing a theory concerned with aviation.

In all cases the model refers to, and highlights, a number of important aspects in theory and in reality. The theory and the model do not represent all that there is in the world, but lay down an ideal system in which a number of important factors operate according to certain game-rules. For example, in the law of free fall of bodies there is no reference to the shape or color of the body, or to the material it is made of. The law works in the framework of a closed and ideal system, in which gravity acts on a huge body or a minute body; in Newton's law of gravity the mass of every body (such as the sun and Earth) is local, as if it is concentrated at one point; and the computer model emphasizes mainly the following properties in the mind: representation of the world by means of symbols on which computational rules act.

Many models are based on analogy. All cognitive psychology is based on a broad analogy, where the relation between mind and brain is like the relation between software and hardware. The analogy makes it possible to use an existing and understood theory that explains a set of computational phenomena in order to explain a set of not-understood mental phenomena. We say that if the mind is like the software, and if the brain is like the hardware, then mental behavior can be explained similarly to how we explain the behavior of the computer.

An important part of the theoretical discussions on the subject under consideration centers on this question: what is the nature of the scientific representation? Giere (2004) suggests that models are designed so that certain elements in the world are represented by the model to realize various aims, and the most important of them is explanation. That is, scientists are they who design models to bear similar features to phenomena that stimulate their curiosity. Suarez (2004) suggests that one of the important properties of scientific representation is the ability to deduce conclusions from the representation to the represented, where drawing these conclusions is practical in essence and need not be accomplished solely by standard deductive methods. Woody (2004) stresses the intimate connection between a scientific representation and explanation, and van Fraassen (2000, 2004) asserts that scientific representations are intentional, created from a certain point of view, and allow explanations to be given to certain aspects in the world. van Fraassen (2000), who suggests seeing Aristotle's two books, the *Poetics* and the *Physics*, as attempts to grasp tragedy and science as different kinds of representations, maintains that representations realize certain functions that are determined by the scientist or the dramaturge. Among other things he suggests that representation in cognitive psychology is metaphorical and analogical in nature, and states in general:

> Science is representation; … there is no representation except representation by someone, through the use of something, to someone. The locus of representation is consciousness, so to speak: a shadow or reflection in the water is not in and of itself a representation. (p. 54)

Mechanisms: A large part of the explanations suggested by scientists for complex phenomena are not based on the use of laws or theories by which means a large number of phenomena can be covered, but by the use of mechanisms, that is, by a description of the general activity of a given system made up of a number of components placed in a certain interaction relationship, interactions that in the end create the general investigated behavior (see discussion in chapter 6). As an example of explanation by means of mechanisms we shall look at the action of the blood circulation system, whose function is to distribute oxygen and calories to the body tissues. This system breaks down into components and sub-components, each of which is characterized by its own activity, for example, the heart, which pumps and compresses the blood, and the kidneys, which cleanse the blood, and several more sub-systems, so that the action of all these parts together ultimately realizes the function of the blood circulation system (see Carver, 2001).

Mechanisms in the natural sciences are largely based on the relation between higher actions and lower actions. The output of the system as a whole is perceived as activity on a high level, while the activity of the parts of the system is perceived as activity on a lower level. Each and every part in the overall system can itself be broken down further to sub-parts, whose activity is perceived as activity on a still lower level than the activity of the parts. That is, the parts of the system are grasped as possessing lower and more basic activity than the action of the system itself; the breakdown is from the whole to the parts that constitute it.

What is the relation between these three approaches and multi-explanation theory? The answer is this: the theory can be analyzed from several viewpoints and it can be illustrated that the theory is anchored to the three approaches to constructing a theory. Evidently, the basic division of a theory according to the received view into a theoretical part, an observational part, and a part that connects these two is maintained in multi-explanation theory too. Without commitment to the positivist approach (e.g., the sharp distinction between observational and theoretical concepts), this theory may be analyzed to a level of theoretical structures and processes based on mental, cognitive, neurophysiological concepts; to a level that deals with the description and measurement of behavioral observations; and to the level that handles the connection between the observations and their representations at the input (stimulus) stage, the output (response) stage, and sometimes even in an attempt to find behavioral indicators of the theoretical concepts that mediate between input and output.

Multi-explanation theory may also be perceived as a procedure for constructing representational models that attempt to reflect real behavioral processes, and to suggest that these processes, described in models, express behavioral processes, hence to propose explanations for the observations. These models are suggested from two main viewpoints: mechanistic explanation models that propose mechanistic explanations for a given behavior, are founded on an objective viewpoint accepted in science that was discussed in earlier chapters. By contrast, mentalistic explanations are proposed from a mixed, objective-subjective, viewpoint. On the one hand, mentalistic explanation schemes are objective, meeting the scientific methodological requirements, and every researcher can use them on the same level of efficiency. On the other hand, these schemes refer to mental states and processes (will, belief, knowledge, consciousness), which are in the domain of the individual alone, that is, states and processes that belong to the subjective domain.

Multi-explanation theory may also be seen as based on giving explanations by means of a description of a behavioral mechanism, a description that rests on the decomposition of the studied behavior into its elementary components (and see chapter 6 on the distinction between decomposition of a mechanistic and a mentalistic system). As decomposition is an extremely important element in multi-explanation theory, I shall expand a little on this matter and raise some general questions:

Is there a universal criterion whereby an entire system can be broken down into its components? Is there a criterion that distinguishes one component from another? Can each and every component be broken down to its sub-components? When does this decomposition process stop? And by what law or rule is it possible to make a synthesis of all the components, to put them together again, and to obtain the phenomenon whole? Is there any universal rule of connection, synthesis, or is the rule of connection context-dependent?

All these are ponderous questions, which have irked philosophers and scholars in the various areas of science, from the era when the Greek philosophers believed that matter could be broken down into atoms, which constituted the most basic materials, to the idea that it is possible break cognitive behavior into sub-components and down to the most primitive parts – those that can express their simple action mechanically (e.g., definition of the binary state by means of two electrical states). Considering this, and the examples I referred to in the previous chapters, for example, decomposition of Max's behavior, the action of the flashlight, and the action of the blood circulation system, my answer to these questions is this. To the best of my understanding this set of questions does not have clear-cut answers, a general formula for disassembly and assembly (analysis and synthesis) suited to different systems, *because the answers depend on the level of scientific knowledge whence the scientist starts out to investigate the given phenomenon.* To substantiate this answer, we shall look at the following hypothetical example: how do different people, at different times in history, explain the action of a simple transistor (from which, for simplicity, all writing has been erased).

1) *The shaman*: How will the medicine-man in a tribe living somewhere in the world, thousands of years ago, understand the working of the transistor? In the morning the shaman wakes up in his hut (or cave) and there before him lies a transistor. After recovering from his surprise, the shaman takes the transistor in his hands and discovers that it can be divided into the following parts: a handle that he can move, an aerial that he can raise, a station search wheel that he can turn, and a light that goes on when he presses the right button. And the moment the transistor light goes on, the shaman will be filled with profound wonder, and will think that this object is an expression of a supreme power that has chosen him, the shaman, as his representative on earth. Here the investigation by the medicine-man will cease and faith will enter. The man will explain the action of the holy transistor through his god's power.
2) *The "Archimedean"*: How will a philosopher from the time of Archimedes understand the action of the transistor? Free of strong religious influence, the Archimedean philosopher is likely to reach the conclusion that the instrument lights up

only when the two cylinders (batteries), located inside the transistor, are arrayed in a certain order, and when the right button is pressed. The man, therefore, will explain the action of the transistor by means of these two necessary conditions, without which the strange instrument does not light up.

3) *The "Faradian"*: How will the scholar of the time of Faraday, the renowned British researcher of electricity, understand the transistor's action? He will undoubtedly arrive at what the Archimedean researcher arrived at – but even more. He will understand that the two cylinders inside the transistor are electric batteries, and after some effort he will even be able to replace the two original batteries with a battery that he himself has made in his laboratory, so that the transistor can continue to be lit up for days on end, even after the two original batteries are exhausted. The researcher will suggest, therefore, a partial explanation of the action of the transistor (illumination) in reference to the theory of electricity prevailing in his day.
4) *The teenage youth*: How will a present-day teenager understand the action of the transistor? Apart from the obvious fact that the youth will, in the end, toss this transistor into the garbage because no name of a prestigious manufacturer appears on it, it is reasonable to suppose that the youth can provide us a first fairly successful explanation of the transistor's action, and even succeed in picking up some (ear splitting) modern metal rock music. Moreover, if by chance the young person is also a geek who likes science, it is very probable that he will be able to give us a quite good scientific explanation of the action of the transistor, to do with radio waves and their reception, and he may even mention Maxwell's equations, Hertz's experiment, Marconi's realization, and the invention of the transistor.

So what does this transistor story tell us about the problem of decomposition into components? Three main things. First, the explanation of the working of the transistor clearly depends on the level of scientific knowledge in the given period. It is hard to imagine that the shaman, the Archimedean, and even the Faradian would be likely to arrive at the proper explanation of the transistor's action, that is, to suggest a suitable breakdown of the transistor into components and with the help of their description answer the question why this instrument lights up, or the question what is its function – an explanation that the teenage boy will have no trouble giving.

Secondly, the knowledge prevailing at that time constitutes the infrastructure for the advance of science, but this information is liable to stymie proper progress also. Without the shaman freeing himself of the strong faith in the superior power that chose him, that is, without this medicine-man (the scientist of his era) releasing himself from the knowledge that controls him and the members of his tribe, it is hard to see how he would dare even to open the battery chamber and reach the level of knowledge that the Archimedean reached, because this gadget is holy.

Thirdly, the first three explanations (the shaman's the Archimedean's, and the Faradian's) may be seen as a chain of explanations based on the methodology of decomposition, which gradually approach the explanation by the teenager, that is, the explanation of the transistor's action as we know it today. The shaman is at the start of the path, and his explanation, even though he tries to base it on a supreme power, is flawed in many respects. For example, his explanation cannot be put to an empirical test; after him the Archimedean makes an enormous leap forward, explaining the action of the transistor by means of two necessary conditions; the Faradian adds to this explanation a partial specification of some of the transistor's components: electricity and batteries; and finally the youth is able to provide the explanation accepted in our time.

9.3 Multi-explanation theory, understanding, explanation, and emergent properties

Several researchers have tried to comprehend the connection between explanation and understanding: understanding is not connected to subjective feeling, to a sense of confidence, to the psychological belief that the researcher has succeeded in his attempts to draw aside the curtain over the secret of nature, but is linked to scientific research, to the properties of the theory, and chiefly to the success of the explanation (see discussions in De Regt, 2004; Salmon, 1998; Trout, 2002). Trout not only argues that the feeling of subjective understanding causes grave mistakes, but that epistemologically understanding does not offer clues to successful explanation. Instead he suggests an objective criterion – precision in explanation, which brings with it valid understanding. De Regt criticizes Trout in several respects and proposes a practical criterion of scientific understanding linked to the properties of the theory (consistency, simplicity, fruitfulness, visibility, etc.) and to its practical use.

Multi-explanation theory proposes understanding not by means of reliance on a subjective feeling of understanding but mainly by reliance on objective criteria linked to the application of mechanistic and mentalistic explanation schemes. In this respect the accuracy in prediction of this theory is increased, because the explanation that the theory suggests is not limited to the area of the mechanistic explanation alone but extends to the mentalistic explanation too. Still, it should be stressed yet again that the theory cannot offer a complete explanation for a complex behavior because the connection between the mental and the physical is not understood.

Salmon suggests that scientific understanding is achieved in two ways: first, by fitting the investigated phenomenon into a comprehensible and overall world picture; and second, by revealing the process, the internal machinery, of the phenomenon:

> … there are at least two intellectual benefits that scientific explanations can confer upon us, namely (1) a unified world picture and insight into how various phenomena fit into that overall scheme, and (2) knowledge of how things in the world work, that is, of the mechanisms, often hidden, that produce the phenomena we want to understand. (p. 89)

I shall call the first way of understanding "scheme fitting" and second way "production mechanism". Multi-explanation theory uses both these kinds of understanding. However, while mechanistic explanations are amenable to classification into both kinds, mentalistic explanations are classifiable only into one kind.

As an example of a mechanistic explanation by means of scheme fitting, I shall note the matching law, which I discussed in earlier chapters: by employing this law learning phenomena are explained by means of their match with the law as a particular case; and as an example of a mechanistic explanation by means of production mechanism I shall note the dualist theory of memory: by means of this theory the creation of memory output is explained through detailing the inner mechanism of memory (according to the analogy of the computer) as based on two stores (short- and long-term) and a few rules of computation that operate on the relevant information that is coded, stored in, and retrieved from these stores. (In the latter example I of course disregarded the mentalistic-consciousness component of short-term memory, and referred to it analogously to computer programs: a limited capacity which deals with a certain kind of information, etc. It is worth noting that most theories in cognitive psychology are of the production mechanism kind, whether these theories are based on classic computer programs or on neural networks.)

In contrast to the mechanistic explanation, the mentalistic explanation scheme can be taken as production mechanism, which details the mental components, for example, will, belief, action, and their relations founded on consciousness and rationality. Salmon (1998) calls understanding of this kind "goal-oriented understanding" (p. 8) and ties it to giving explanations by an appeal to conscious motivations and goals. (He distinguishes explanation of this kind from functional explanation, based ultimately on a causal or survival interpretation.) This mentalistic scheme is applied to a certain individual (human, animal) and describes how the mentalistic mechanism produces the specific behavior under research. The mentalistic explanation does not offer us scientific understanding by scheme fitting (by fitting to a law), but by production mechanism, which in the present case is a mental mechanism or process.

In the introduction to his book, and in accordance with the article on understanding included in the book, Salmon (1998) suggests that these two kinds of explanation (by scheme fitting and by production mechanism) are not contradictory but complement each other: basic production mechanism is, in his opinion, also the discovery of a unifying principle in nature. In support of this proposal, Salmon presents an example from the world of physics (movement of a balloon in a rising airplane), which he believes is well explained by both production mechanism and scheme fitting. However, I would like to present a somewhat different point of view.

I wish to end this section with a simple illustration, the mosaic example, which suggests that these two kinds of explanation are independent, even though they have to be combined in order to propose an explanation for the generation of the mosaic example (see figure 9.1 and below). Furthermore, as we shall see later, this example has an interesting methodological implication for the connection between the perception that conscious behavior possesses emergent properties and multi-explanation theory.

9.3.1 Two kinds of explanation (scheme fitting, production mechanism) and the mosaic example

Figure 9.3.1 The mosaic example. (In this example I have highlighted several mosaic patterns by thickening them. For explanations see text.)

On seeing this mosaic the question at once arises, how is the mosaic created? The answer is rather simple. The mosaic is created by a simple program (written according to my instructions by my research assistant Ms. Irena Polovso). Without going into a detailed description of this program, it is possible to characterize it in the following way: the mosaic program contains four different shapes, an 11x11 pigeonhole board, and some probability rules which determine the placing of the shapes on the board, one shape after another, each shape in the space next to the space in which the previous shape was placed. Rule (a) allows determination of probability of placement of each shape (probabilities must add up to 1); rule (b) allows determination of the probability of placing a shape next to a shape on the board, or placing a shape in a pigeonhole on the corner of a pigeonhole in which a shape has been placed; rule (c) allows determination of the conditional probabilities of the appearance of a shape given a shape on the board. These probabilities are determined before the running of the program, which begins with placing a shape in the middle of the board. In the example in figure 9.3.1, for simplicity I used two shapes only: a circle and a solid square (i.e., we ascribed to the other shapes probability of zero). In like manner, therefore, it is possible to create a lot of mosaics of this kind or of any other kind with two, three, and four shapes, where the creation of all these examples will have the same explanation of how they were created: by running the mosaic program!

However, examination of the mosaic reveals something new: assemblies of squares and circles create several interesting and entirely new formations, "mosaic patterns": triangles, squares, rectangles, straight and sinuous lines, loops of one color that surround shapes of another color, and symmetrical relations of between different shapes. (This perception of mosaic patterns is like seeing a human or an animal face in the clouds.)

The question that arises following this scrutiny is the following: is it possible to explain these mosaic patterns by means of the mosaic program in itself? The answer is no.

First, to suggest an explanation for the appearance of mosaic patterns (to show the stages of building a given mosaic pattern and to calculate the probability of its appearance) we must first develop a system of geometrical concepts, a theory of "geometrical schemes", by means of which it is possible to identify the mosaic patterns. Without our having a system of geometrical schemes, parts in the mosaic cannot be sorted into such patterns, and it cannot be stated, for example, that this assembly of solid squares is an unbroken loop that encompasses two circles; or that these six squares form the pattern of a rectangle. Hence, the explanation of why the collection of shapes in the mosaic has a special and particular pattern does not depend on *production mechanism* (the mosaic program) but on *scheme fitting* (geometrical schemes).

Secondly, it is hard to see how on the basis of the mosaic program itself the geometrical-scheme system can be developed. This is a system of concepts that was developed independently of the mosaic program.

Before us, then, is an example where on the one hand there is independence of the two kinds of explanation: production mechanism (the mosaic program) and scheme fitting (geometrical schemes), and on the other hand the combination of these two explanations is able to offer a sound explanation for a given mosaic pattern: how it is created and what is the probability of its appearance.

9.3.2 Emergent properties and the mosaic example

Several researchers maintain that conscious behavior possesses emergent properties defined as follows:

> Emergent properties are irreducible to, and unpredictable from, the lower-level phenomena from which they emerge. (Kim, 1966, p. 228)

Kim goes on to clarify the term reduction from the viewpoint of a researcher who believes in emergent properties, and suggests an explanatory interpretation of this concept. Following him, I suggest that emergent properties (which appear in new phenomena I call "emergent phenomena") are new properties that do not appear in the source phenomenon from which they sprang, and it is hard to explain them by an appeal to the properties of the source phenomenon. For example, as we saw in the earlier chapters (especially chapter 8), no theory that explains consciousness has yet been found, notwithstanding the accepted opinion that consciousness is associated with the activity of the brain; likewise, it is hard to understand the experience of pain through recourse to the action of the special kind of nerves called C-fibers. Consciousness, feeling pain, and other mentalistic phenomena are therefore considered emergent phenomena: these are phenomena possessing new qualities that are not found in the source phenomenon, and a way has not yet been found to explain them, not by production mechanism, that is, by finding the neurophysiological mechanism that produces consciousness, nor by scheme fitting, that is, by reducing mental concepts to neurophysiological terms.

Mental behavior is not the only emergent phenomenon known to us. Science teaches us that not only are there other emergent phenomena but that at least for some of them a satisfactory explanation is found. May something be learnt from these cases that will help us understand consciousness as an emergent phenomenon? As possible answers, I shall examine two cases in which emergent phenomena have received satisfactory explanations: chemical reactions and the mosaic pattern.

Chemical reactions: The history of the natural sciences teaches us that in several cases they have been able to offer explanations for many phenomena that were thought to possess emergent properties. We shall look at the example of chemical reactions. Several researchers, following the British tradition of science and philosophy, proposed that these new properties cannot be explained on the basis of the properties of the source phenomenon. For example, water, which is created by the fusion of two gases, oxygen and hydrogen, constitutes two phenomena on different levels of description. But this perception changed entirely with the development of quantum theory, which could explain chemical processes on the basis of sub-atomic processes (see discussion on this matter in McLaughlin, 1992. On emergent phenomena see collection of articles in Beckermann, Flohr & Kim, 1992).

Is it possible to suggests, similarly, a psycho-physical theory that explains consciousness and the brain/consciousness relation, based on quantum theory? Several articles proposing theories of this kind (see, e.g., Hameroff & Penrose, 1997) are included in the collection of articles edited by Shear (1997) on the 'hard problem' developed by Chalmers (1996, 1997a). In a paper in response to those articles, Chalmers (1997b) writes that works founded on quantum theory are still at the speculative level, and it is not clear how the structure of the description and the quantum explanation are linked to the structure of consciousness.

The mosaic example: Is the mosaic example likely to teach us anything relevant to the explanation of consciousness as emergent phenomena? The answer to this question depends on the answer to the following one: can mosaic patterns be taken as possessing emergent properties that are given a satisfactory explanation? To answer this latter question I shall first summarize the three main and most relevant characteristics for our concern with this example:

(a) By means of a system of geometrical schemes it is possible to identify in the mosaic example different mosaic patterns. These mosaic patterns differ conceptually from an accidental group of several shapes in the mosaic example. These are patterns possessing new properties, possessing new meanings imparted to them by the system of geometrical schemes.

(b) By means of the mosaic program it is not possible to explain the nature of mosaic patterns or the system of the geometrical schemes whereby mosaic patterns are identified. Furthermore, identifying mosaic patterns has no influence on the production of the mosaic program. (The latter point has an interesting implication for "downward causation", according to which emergent phenomena have a causal influence on the source phenomenon. This is an elaborate subject, which is beyond the present purpose, so I shall not expand on it. See, e.g., Kim, 1992, 1996.)

(c) Given an identified mosaic pattern, by means of the mosaic program it is possible to explain the stages of construction of this mosaic pattern and to calculate the probability of its occurrence.

In light of these characteristics it comes out, first, that a given mosaic pattern possesses an emergent property according to (a) and (b), and second, that a satisfactory explanation for this pattern may be offered according to characteristic (c).

Given this answer, I shall return to my previous question: can the mosaic example teach us anything relevant to the explanation of Conscious behavior as emergent phenomena? That is, is it possible to explain mentalistic phenomena as we explain the mosaic patterns? The answer depends on the following analogies:

- between the individual's brain and the mosaic program, where the brain, like the mosaic program, creates a behavior mosaic;
- between the consciousness system and the system of geometrical schemes, where a part of the behavior mosaic acquires a special characteristic, a new property, by the mental system, and as a result turns into a mental behavior pattern (perhaps by conscious-induction, see chapter 8).

On the assumption that these analogies indeed hold, it may be possible to suggest according to characteristic (c) a guideline to explain conscious behavior as an emergent phenomenon: given a mental behavior pattern, we shall explain by means of cerebral processes the stages of construction of this mental behavior pattern and we shall calculate the probability of its appearance.

However, it seems to me that this analogy rests on shaky ground. First, given a mental behavior pattern, it is hard to see how this behavior pattern can be built by an appeal to neurophysiological knowledge of the brain, or how the probability of its occurrence can be calculated by resorting to this knowledge. (Clearly, the production of a given behavior involves other complicated mechanisms.) While in the case of the mosaic example it is easy to offer these explanations for any (identified) given mosaic pattern, to the best of my knowledge (and see discussions in the previous chapters, especially chapter eight), we do not have the means to offer an explanation of this kind for mental behavior. For example, it is hard to see how it is possible to explain the components of the following mental behavior pattern (i.e., to describe how the components were created and developed, and to calculate the probability of their occurrence): David's desire to meet Ruth in Tel Aviv, his belief that a journey by train is the most efficient means of realizing this desire, and the action that realizes his desire.

Secondly, an analogy may be suggested between a geometrical-scheme system, which identifies different mosaic patterns, and the mental system, which makes a certain behavior in the mosaic of behaviors a behavior pattern carrying conscious meaning. However, this analogy is nothing but a very loose similar feature. It is quite clear how to identify mosaic patterns by means of a system of geometric schemes (by a process of scheme fitting); but we do not have a theory or process that clarifies for us how the mental system makes a behavior a mental behavior pattern. All we know is that it is possible to suggest a practical, mentalistic explanation for a given behavior by an appeal to internal conscious factors, for example, an appeal to the individual's will/belief.

Thirdly, a system of geometrical schemes is external to and independent of the mosaic program; but the consciousness system depends on the brain, which in a way not understood is responsible for its creation; and while it is very easy to explain how a system of geometrical schemes is formed, there is still no theory explaining how the consciousness system is created and functions.

What is the implication of this discussion for multi-explanation theory? To my mind the conclusion arising from this discussion is as follows: although it is accepted to think that conscious behavior originates in the action of the brain, it does not seem possible to suggest an explanation for consciousness as an emergent phenomenon, not by analogy to chemical reactions and not by analogy to the mosaic example. Methodologically, then, it is useful to use multi-explanation theory as a means of explaining mental behavior because this theory can deal coherently with the two main and important components of this complex behavior: on the one hand it can deal with the neurophysiological-behavioral system by using mechanistic explanation schemes, and on the other hand it can deal with the mental-consciousness system by using mentalistic explanation schemes, schemes that have undergone Scientification, so they are proper from the viewpoint of scientific methodology.

References

Adams, D. B. (1979). Brain mechanisms for offense, defense and submission. *Behavioral and Brain Sciences,* 2, 201–241.

Alcock, J. (1998). *Animal behavior: An evolutionary approach.* Sunderland, MA: Sinauer Associates.

Allen, C. (1992). Mental content. *The British Journal for the Philosophy of Science,* 43, 537–553.

Allen, C. (1995). Intentionality: Natural and artificial. In H. L. Roitblat & Jean-Arcady Meyer (Eds.), *Comparative approaches to cognitive science.* Cambridge, MA: The MIT Press.

Allen, C. (1997). Animal cognition and animal minds. In M. Carrier & P. K. Machamer (Eds.), *Mindscapes:Philosophy, science, and the mind.* Pittsburgh: Pittsburgh University Press/Germany: UVK. Universitatsverlag Konstanz.

Allen, C. & Bekoff, M. (1995). Cognitive ethology and the intentionality of animal behaviour. *Mind & Language,* 10, 313–328.

Allen, C. & Bekoff, M. (1997). *Species of mind: The philosophy and biology of cognitive ethology.* Cambridge, MA: The MIT Press.

Anderson, P. W. (1972). More is different: Broken symmetry and the nature of the hierarchical structure of science. *Science,* 177, 393–396.

Audi, R. (1993). *Action, intention, and reason.* Ithaca: Cornell University Press.

Baddeley, A. D. (1976). *The psychology of memory.* New York: Basic Books.

Baerends, G. P. (1976). The functional organization of behavior. *Animal Behaviour,* 24, 726–738.

Bandura, A. (1986). *Social foundations of thought and action: A social cognitive theory.* Englewood Cliffs, NJ: Prentice Hall.

Barendregt, M. (2003). Genetic explanation in psychology. *The Journal of Mind and Behavior,* 24, 67–90.

Barendregt, M. & van Rappard, H. (2004). Reductionism revisited: On the role of reduction in psychology. *Theory & Psychology,* 14, 453–474.

Barnett, S. A. (1998). Instinct. In G. Greenberg & M. M. Haraway (Eds.), *Comparative psychology: A handbook* (pp.138–149). New York: Garland Publishers.

Barrett, P. & Bateson, P. (1978). The development of play in cats. *Behaviour,* 66, 106–120.

Bateson, P. & Young, M. (1981). Separation from the mother and the development of play in cats. *Animal Behaviour,* 29, 173–180.

Bechtel, W. & Abrahamsen, A. (1991). *Connectionism and the mind.* Cambridge, MA: Basil Blackwell.

Bechtel, W. & Mundale, J. (1999). Multiple realizability revisited: Linking cognitive and neural states. *Philosophy of Science,* 66, 175–207.

Bechtel, W. & Richardson, R. C. (1993). *Discovering complexity: Decomposition and localization as strategies in scientific research.* Princeton, NJ: Princeton University Press.

Beckermann, A., Flohr, H. & Kim, J. (Eds.) (1992). *Emergence or reduction? Essays on the prospects of nonreductive physicalism.* Berlin: Walter de Gruyter

Bekoff, M. (2002). *Minding animals: Awareness, emotions, and heart.* New York: Oxford University Press.

Bekoff, M. & Allen, C. (1997). Cognitive ethology: Slayers, skeptics, and proponents. In R. W. Mitchell, N.S. Thompson & H.L. Miles (Eds.), *Anthropomorphism, anecdotes, and animals* (pp. 313–334). New York: State University of New York Press.

Bem, S. & Looren de Jong, H. L. (1997). *Theoretical issues in psychology: An introduction.* London: Sage.

Benjafield, J. G. (1997). *Cognition* (2nd ed.). New Jersey: Prentice Hall.

Bennett, J. (1991). How is cognitive ethology possible? In C. A. Ristau (Ed.), *Cognitive ethology: The minds of other animals.* Hillsdale, NJ: LEA.

Ben-Zeev, A. (1993). *The perceptual system: A philosophical and psychological perspective.* New York: Peter Lang.

Bickle, J. (1998). *Psychoneural reduction: The new wave.* Cambridge, MA: The MIT Press.

Bird, A. (2000). *Philosophy of science.* London: Routledge.

Block, N. (1978/1980). Troubles with functionalism. In N. Block (Ed.), *Readings in philosophy of psychology* (vol. 1). Cambridge, MA: Harvard University Press.

Block, N. (1995). The mind as the software of the brain. In E. E. Smith & D. N. Osherson (Eds.), *Thinking: An invitation to cognitive science* (2nd ed., vol. 3). Cambridge, MA: The MIT Press.

Block, N. & Alston, W. P. (1984). Psychology and philosophy. In M. H. Bernstein (Ed.), *Psychology and its allied disciplines.* New Jersey: LEA.

Bontly, T. (2000). Review of John Bickle's *Psychoneural reduction: The new wave. British Journal for the Philosophy of Science,* 51, 901–905.

Boring, E. G. (1950). *A history of experimental psychology* (2nd ed). New York: Appleton-Century-Crofts.

Boyd, R. & Richerson, P. J. (1988). An evolutionary model of social learning: The effect of spatial and temporal variation. In T. R. Zentall & B. G. Galef, Jr. (Eds.), *Social learning: Psychological and biological perspectives* (pp. 29–48). Hillsdale, NJ: Erlbaum.

Bradshaw, J. W. S. (2002). *The behaviour of the domestic cat.* New York: CABI Publishing.

Bradshaw, J. W. S., Healey, L. M., Thorne, C. J., Macdonald, D. W. & Arden-Clark, C. (2000). Differences in food preferences between individuals and populations of cats. *Applied Animal Behaviour Science,* 68, 257–268.

Breznitz, S. (1989). Information-induced stress in humans. In S. Breznitz & O. Zinder (Eds.), *Molecular biology of stress.* New York: Liss.

Bromley, D. B. (1986). *The case-study method in psychology and related disciplines.* New York: Wiley.

Brook, A. & Stainton, R. J. (2001). *Knowledge and mind: A philosophical introduction.* Cambridge, MA: The MIT Press.

Bunge, M. (1997). Mechanism and explanation. *Philosophy of the Social Sciences,* 27, 410–465.

Burghardt, G. M. (1985). Animal awareness: Current perceptions and historical perspective. *American Psychologist,* 40, 905–919.

Burghardt, G. M. (1991). Cognitive ethology and critical anthropomorphism: A snake with two heads and hognose snakes that play dead. In C.A Ristau (Ed.), *Cognitive ethology: The minds of other animals* (pp. 53–90). Hillsdale, NJ: LEA.

Burghardt, G. M. (1998). Play. In G. Greenberg & M. M. Haraway (Eds.), *Comparative psychology: A handbook* (pp. 725–735). New York: Garland.

Campbell, N. R. (1953). *What is science?* New York: Dover.

Caro, T.M. (1981). Predatory behaviour and social play in kittens. *Behaviour,* 76, 1–24.

Cartwright, N. (1999). *The dappled world: A study of the boundaries of science.* Cambridge: Cambridge University Press.

Carver, C. F. (2001). Role functions, mechanisms, and hierarchy. *Philosophy of Science,* 68, 53–74.

Carver, C. F. (2002). Structures of scientific theories. In P. Machamer & M. Silberstein (Eds.), *The Blackwell guide to the philosophy of science.* Malden, MA: Blackwell.

Causey, R. J. (1972). *Unity of science.* Dordrecht, Holland: D. Reidel.

Chalmers, D. J. (1996). *The conscious mind: In search of a fundamental theory.* New York: Oxford University Press.

Chalmers, D. J. (1997a). Facing up to the problem of consciousness. In J. Shear (Ed.), *Explaining consciousness – the 'hard problem'.* Cambridge, MA: The MIT Press.

Chalmers, D. J. (1997b). Moving forward on the problem of consciousness. In J. Shear (Ed.), *Explaining consciousness – the 'hard problem'.* Cambridge, MA: The MIT Press.

Chalmers, D. J. (2003). Consciousness and its place in nature. In S. P. Stich & T. A. Warfield (Eds.), *The Blackwell guide to philosophy of mind* (pp. 102–142). Malden, MA: Blackwell.

Churchland, P. M. (1981). Eliminative materialism and the propositional attitudes. *Journal of Philosophy,* 78, 67–90.

Churchland, P. M. (1988). *Matter and Consciousness* (revised edition). Cambridge, MA: The MIT Press.

Churchland, P. M. (1989). Folk psychology and the explanation of human behavior. In P. M. Churchland (Ed.), *A neurocomputational perspective* (pp. 11–127). Cambridge, MA: The MIT Press.

Clarke, R. (2003). Freedom of the will. In S. P. Stich & T. A. Warfield (Eds.), *The Blackwell guide to philosophy of mind* (pp. 369–404). Malden, MA: Blackwell Publishing.

Coombs, C. H., Dawes, R. M. & Tversky, A. (1970). *Mathematical psychology: An elementary introduction.* Englewood Cliffs, NJ: Prentice Hall.

Copeland, B. J. (1993). *Artificial intelligence.* Oxford: Blackwell.

Coren, S. & Girus, J. S. (1978). *Seeing is deceiving: The psychology of visual illusions.* Hillsdale, NJ: LEA.

Crane, T. (1991). *The mechanical mind: A philosophical introduction to minds, machines and mental representation.* London: Penguin Books.

Crist, E. (1999). *Images of animals: Anthropomorphism and animal mind.* Philadelphia: Temple University Press.

Cronbach, L. J. (1975). Beyond the two disciplines of scientific psychology. *American Psychologist,* 30, 116–127.

Cummins, R. (1983). *The nature of psychological explanation.* Cambridge, MA: The MIT Press.

Cummins, R. (2000). "How does it work?" versus "What are the laws?" Two conceptions of psychological explanation. In F. C. Keil & R. A. Wilson (Eds.), *Explanation and cognition.* Cambridge, MA: The MIT Press.

Darwin, C. (1871/1982). *The descent of man.* New York: Modern Library.

Davidson, D. (1975). Thought and talk. In S. Guttenplan (Ed.), *Mind and language.* Oxford: Oxford University Press.

Davidson, D. (1980). *Essays on actions and events.* Oxford: Clarendon Press.

Davis, H. (1997). Animal cognition versus animal thinking: The anthropomorphic error. In R. W. Mitchell, N. S. Thompson & H. L. Miles (Eds.), *Anthropomorphism, anecdotes, and animals* (pp. 335–347). New York: State University of New York Press.

Davison, M. & McCarthy, D. (1988). *The matching law: A research review.* Hillsdale, NJ: LEA.

Dawkins, M. S. (1995). *Unravelling animal behavior* (2nd ed.). Essex: Longman Scientific & Technical.

De Regt, H. W. (2004). Discussion note: Making sense of understanding. *Philosophy of Science,* 71, 98–109.

de Waal, F. B. M. (1997a). Foreword. In R. W. Mitchell, N. S. Thompson & H. L. Miles (Eds.), *Anthropomorphism, anecdotes, and animals* (pp. xiii-xvi). New York: State University of New York Press.

Deitel, H. M. & Deitel, B. (1985). *Computers data and processing.* New York: Academic Press.

Dennett, D. C. (1969). *Content and consciousness.* London: Routledge & Kegan Paul.

Dennett, D. C. (1971). Intentional systems. *Journal of Philosophy,* 8, 87–106.

Dennett, D. C. (1979). *Brainstorms: Philosophical essays on mind & psychology.* Cambridge, MA: The MIT Press.

Dennett, D. C. (1987). *The intentional stance.* Cambridge, MA: The MIT Press.

Dennett, D. C (1991). *Consciousness explained.* Boston: Little, Brown and Company.

Dennett, D. C. (1995). Do animals have beliefs? In H. L. Roitblat & Jean-Arcady Meyer (Eds.), *Comparative approaches to cognitive science.* Cambridge, MA: The MIT press.

Dietrich, E. & Hardcastle, V. G. (2005). *Sisyphus's boulder: Consciousness and the limits of the knowable.* Amsterdam/Philadelphia: John Benjamins.

Domjan, M. (1998). *The principles of learning and behavior* (4th ed.). Pacific Grove, CA: Brooks/Cole.

Dray, W. H. (1966). *Philosophical analysis and history.* New York: Harper & Row.

Dretske, F. I. (1986). Misrepresentation. In R. J. Bogdan (Ed.), *Belief.* Oxford: Clarendon Press.

Duhem, P. (1996). *Essays in the history and philosophy of science.* Indianapolis and Cambridge: Hackett. (Translated by R. Ariew and P. Baker.)

Edelman, G. M. & Tononi, G. (2000). *A universe of consciousness: How matter becomes imagination.* New York: Basic Books.

Eibl-Eibesfeld, I. (1975). *Ethology: The biology of behavior* (2nd ed.). New York: Holt, Rinehart, and Winston.

Ekstrom, L. W. (2000). *Free will: A philosophical study.* Boulder, CO: Westview Press.

Epstein, R. (1998). Anthropomorphism. In G. Greenberg & M. M. Haraway (Eds.), *Comparative psychology: A handbook* (pp. 71–73). New York: Garland.

Feest, U. (2003). Functional analysis and autonomy of psychology. *Philosophy of Science,* 70, 937–948.

Fetzer, J. H. (2001). *Computers and cognition: Why minds are not machines.* Dordrecht: Kluwer Academic Publishers.

Feynman, R. P. (1985). *QED: the strange theory of light and matter.* New Jersey: Princeton University Press.

Frank, D. (1984). *Behavior of the living: Introduction to ethology.* Tel Aviv: Hakibbutz Hameuchad (in Hebrew).

Flanagan, O. (1992). *Consciousness reconsidered.* Cambridge, MA: The MIT Press.

Flynn, J. P. (1972). Patterning mechanisms, patterned reflexes, and attack behavior in cats. *Nebraska Symposium on Motivation,* 20, 125–154.

Fodor, J. (1974). Special sciences, or the disunity of science as a working hypothesis. *Synthese,* 28, 97–115.

Fodor, J. (1976). *The language of thought.* Sussex: The Harvester Press.

Fodor, J. (1981). *Representations: Philosophical essays on the foundations of cognitive science.* Cambridge, MA: The MIT Press.

Fodor, J. (1987). *Psychosemantics: The problem of meaning in the philosophy of mind.* Cambridge, MA: The MIT Press.

Fodor, J. (1994). *The elm and the expert: Mentalese and its semantics.* Cambridge, MA: The MIT Press.

Fodor, J. (1998). Special sciences: Still autonomous after all these years (a reply to Jaegwon Kim's "M-R and the metaphysics of reduction"). In J. Fodor (Ed.), *In critical condition: Polemical essays on cognitive science and philosophy of mind.* Cambridge, MA: The MIT Press.

Fodor, J. A. & Pylyshyn, Z. (1988). Connectionism and cognitive architecture: A critical analysis. *Cognition,* 28, 3–71.

Franklin, S. P. (1995). *Artificial minds.* Cambridge, MA: The MIT Press.

Frensch, P. A. & Runger, D. (2003). Implicit learning. *Current Directions in Psychological Science,* 12, 13–18.

Gallup, G. Jr. (1998). Animal self-awareness: A debate – can animals empathize? Yes. *Scientific American Presents: Exploring Intelligence,* 9, 66–71.

Gariepy, J. L. (1998). Historical and philosophical foundations of comparative psychology. In G. Greenberg & M. M. Haraway (Eds.), *Comparative psychology: A handbook* (pp. 31–43). New York: Garland.

Giere, R. N. (2004). How models are used to represent reality. *Philosophy of Science,* 71, 742–752.

Ginet, C. (2002). Reasons explanations of action: Causalist versus noncausalist accounts. In R. Kane (Ed.), *The Oxford handbook of free will.* Oxford: Oxford University Press.

Glymour, K. (1980). *Theory and evidence.* Princeton, NJ: Princeton University Press.

Goldstein, R. (2005). *Incompleteness: The proof and paradox of Kurt Gödel.* New York: W. W. Norton.

Gomm, R., Hammersley, M. & Foster, P. (Eds.) (2000). *Case study method: Key issues, key texts.* London: Sage.

Gordon, R. (1986). Folk psychology as simulation theory. *Mind and Language,* 7, 158–171.

Graham, K. (2002). *Practical reasoning in a social world: How we act together.* Cambridge: Cambridge University Press.

Griffin, D. R. (1976). *The question of animal awareness: Evolutionary continuity of mental experience.* New York: Rockefeller University Press.

Griffin, D. R. (1981). *The question of animal awareness: Evolutionary continuity of mental experience* (revised and expanded edition). Los Altos, CA: William Kaufmann.

Griffin, D. R. (2001). *Animal minds: Beyond cognition to consciousness.* Chicago: University of Chicago Press.

Haberlandt, K. (1997). *Cognitive psychology* (2nd ed.). Boston: Allyn and Bacon.

Hall, S. L. & Bradshaw, J. W. S. (1998). The influence of hunger on object play by adult domestic cats. *Applied Animal Behaviour Science,* 58, 143–150.

Hall, S. L., Bradshaw, J. W. S. & Robinson, I. H. (2002). Object play in adult domestic cats: The roles of habituation and disinhibition. *Applied Animal Behaviour Science,* 79, 263–271.

Hamel, J. (1993). *Case study methods.* Newbury Park, CA: Sage.

Hamerroff, S. R. & Penrose, R. (1997) Conscious events as orchestrated space-time selections. In J. Shear (Ed.), *Explaining consciousness – the 'hard problem'.* Cambridge, MA: The MIT Press.

Haraway, M. M. (1998). Species-typical behavior. In G. Greenberg & M. M. Haraway (Eds.), *Comparative psychology: A handbook* (pp. 191–197). New York: Garland.

Harnad, S. (1990). The symbol grounding problem. *Physica D*, 42, 335–346.

Harnad, S. (2000). Correlation vs. causality: How/why the mind-body problem is hard. *Journal of Consciousness Studies*, 7, 54–61.

Heil, J. (1998). *Philosophy of mind: A contemporary introduction.* London: Routledge.

Heil, J. (2003). Mental causation. In S. P. Stich & T. A. Warfield (Eds.), *The Blackwell guide to philosophy of mind* (pp. 214–234). Malden, MA: Blackwell.

Hempel, C. G. (1965). *Aspects of scientific explanation and other essays in the philosophy of science.* New York: The Free Press.

Hempel, C. G. & Oppenheim, P. (1948). Studies in the logic of explanation. *Philosophy of Science*, 15, 135–175.

Herrnstein, R. (1961). Relative and absolute strength of response as a function of frequency of reinforcement. *Journal of the Experimental Analysis of Behavior*, 4, 267–272.

Heyes, C. & Dickenson, A. (1990). The intentionality of animal action. *Mind & Language*, 5, 87–104.

Heyes, C. & Dickenson, A. (1995). Folk psychology won't go away: Response to Allen and Bekoff. *Mind & Language*, 10, 329–332.

Hilgard, E. R. & Bower, G. H. (1966). *Theories of learning* (3rd ed.). New York: Appleton-Century-Crofts.

Hintzman, D. L. (1993). Twenty-five years of learning and memory: Was the cognitive revolution a mistake? In D. E. Meyer & S. Kornblum (Eds.), *Attention and performance XIV* (pp. 359–391). Cambridge, MA: The MIT press.

Hogan, J. A. (1994). The concept of cause in the study of behavior. In J. A. Hogan & J.J. Bolhuis (Eds.), *Causal mechanisms of behavioural development.* Cambridge: Cambridge University Press.

Hogan, J. A. (1998). Motivation. In G. Greenberg & M. M. Haraway (Eds.), *Comparative psychology: A handbook* (pp. 164–175). New York: Garland.

Hon, G. & Rakover, S. S. (Eds.) (2001). *Explanation: Theoretical approaches and applications.* Dordrecht: Kluwer Academic Publishers.

Horgan, T. & Tienson, J. (2002). The intentionality of phenomenology and the phenomenology of intentionality. In D. J. Chalmers (Ed.), *Philosophy of mind: Classical and contemporary readings.* New York: Oxford University Press.

Horgan, T. & Woodward, J. (1985). Folk psychology is here to stay. *The Philosophy Review*, 94, 197–226.

Hoyningen-Huene, P. (2006). Context of discovery versus context of justification and Thomas Kuhn. In J. Schickore & F. Steinle (Eds.), *Revisiting discovery and justification: Historical and philosophical perspectives on the context distinction.* Dordrecht: Springer.

Jackson, F. (1982). Epiphenomenal qualia. *Philosophical Quarterly*, 32, 127–136.

Jackson, F. (1986). What Mary didn't know. *Journal of Philosophy*, 83, 291–295.

Jamieson, D. & Bekoff, M. (1992). On the aims and methods of cognitive ethology. *PSA*, 2, 110–124.

Jensen, (2002). *The ethology of domestic animals: An introductory text.* Wallingford, UK: CABI.

John, E. R., Chesler, P., Bartlett, F. & Victor, I. (1968). Observational learning in cats. *Science*, 159, 1489–1491.

Johnson, R. N. (1972). *Aggression in man and animals.* Philadelphia: W. B. Saunders.

Josephson, J. R. & Josephson, S. G. (Eds.) (1994). *Abductive inference: Computation, philosophy, technology.* Cambridge: Cambridge University Press.

Kane, R. (Ed.) (2002a). *The Oxford handbook of free will.* Oxford: Oxford University Press.

Kane, R. (2002b), Introduction: The contours of contemporary free will debates. In R. Kane (Ed.), *The Oxford handbook of free will*. Oxford: Oxford University Press.

Kant, I. (1790/1964). *The critique of judgment*. Oxford: Clarendon Press.

Keijzer, F. (2001). *Representation and behavior*. Cambridge, MA: The MIT Press.

Kennedy, J. S. (1992). *The new anthropomorphism*. Cambridge: Cambridge University Press.

Kim, J. (1992). "Downward causation" in emergentism and non-reductive physicalism. In A. Beckermann, H. Flohr, & J. Kim (Eds.), *Emergence or reduction? Essays on the prospects of nonreductive physicalism*. Berlin: Walter de Gruyter.

Kim, J. (1993). *Supervenience and mind: Selected philosophical essays*. New York: Cambridge University Press.

Kim, J. (1996). *Philosophy of mind*. Boulder, CO.: Westview Press.

Kim, J. (1998). *Mind in a physical world: An essay on the mind-body problem and mental causation*. Cambridge, MA: The MIT press.

Kimble, G. A. (1961). *Hilgard and Marquis' conditioning and learning* (2nd ed.). New York: Appleton-Century-Crofts.

Kitcher, P. (1984). In defense of intentional psychology. *The Journal of Philosophy*, 81, 89–106.

Kitcher, P. (1989). Explanatory unification and the causal structure of the world. In P. Kitcher & W. Salmon (Eds.), *Scientific explanation* (pp. 410–505). Minneapolis: University of Minnesota Press.

Kripke, S. (1972/1980). *Naming and necessity*. Cambridge, MA: The MIT Press.

Lazarus, R. S. (1991). *Emotion and adaptation*. New York: Oxford University Press.

Lehman, H. (1997a). Anthropomorphism and scientific evidence for animal mental states. In R. W. Mitchell, N. S. Thompson & H. L. Miles (Eds.), *Anthropomorphism, anecdotes, and animals* (pp. 104–115). New York: State University of New York Press.

Lehner, P. N. (1996). *Handbook of ethological methods*. Cambridge: Cambridge University Press.

Lehrman, D. S. (1953). A critique of Konrad Lorenz's theory of instinctive behavior. *Quarterly Review of Biology*, 28, 337–363.

Lehrman, D. S. (1970). Semantic and conceptual issues in the nature-nurture problem. In L. R. Aronson, E. Tobach, D. S. Lehrman & J. S. Rosenblatt (Eds.), *Development and evolution of behavior: Essays in memory of T.C. Shneirla*. San Francisco: W. H. Freeman.

Leibowitz, Y. (1982). *Body and soul: The psycho-physical problem*. Tel Aviv: Ministry of Defense Press (in Hebrew).

Levi, Z. & Levi, N. (2002). *Ethics, feelings, and animals: The moral status of animals*. Tel Aviv: Sifriyat Hapoalim and Haifa University Press (in Hebrew).

Levine, J. (1983). Materialism and qualia: The explanatory gap. *Pacific Philosophical Quarterly*, 64, 354–361.

Leyhausen, P. (1979). *Cat behavior: The predatory and social behavior of domestic and wild cats*. New York: Garland STPM press.

Libet, B. (2002). Do we have free will? In R. Kane (Ed.), *The Oxford handbook of free will*. Oxford: Oxford University Press.

Lipton, P. (1991). *Inference to the best explanation*. London: Routledge.

Lipton, P. (1992). The seductive-nomological model. *Studies in History and Philosophy of Science*, 23, 691–698.

Lipton, P. (2001a). What good is an explanation? In G. Hon & S.S. Rakover (Eds.), *Explanation: Theoretical approaches and applications* (pp. 43–59). The Netherlands: Kluwer Academic Publishers.

Lipton, P. (2001b). Is explanation a guide to inference? A reply to Wesley C. Salmon. In G. Hon & S.S. Rakover (Eds.), *Explanation: Theoretical approaches and applications* (pp. 91–203). The Netherlands: Kluwer Academic Publishers.

Looren de Jong, H. (2003). Causal and functional explanations. *Theory & Psychology*, 13, 291–317.

Lorenz, K. Z. (1950). The comparative method in studying innate behaviour patterns. *Symposia of the Society for Experimental Biology*. Berlin: Springer-Verlag.

Lorenz, K. Z. (1965). *Evolution and modification of behavior*. London: Methuen.

Losee, J. (1993). *A historical introduction to the philosophy of science* (3rd ed.). Oxford: Oxford University Press.

Ludwig, K. (2003). The mind-body problem: An overview. In S. P. Stich & T. A. Warfield (Eds.), *The Blackwell guide to philosophy of mind* (pp. 1–46). Malden, MA: Blackwell.

Lycan, W. G. (2003). The mind-body problem. In S. P. Stich & T. A. Warfield (Eds.), *The Blackwell guide to philosophy of mind* (pp. 47–64). Malden, MA: Blackwell.

Machamer, P., Darden, L. & Carver, C. (2000). Thinking about mechanisms. *Philosophy of Science*, 67, 1–25.

Mackintosh, N. J. (1974). *The psychology of animal learning*. New York: Academic Press.

Macmillan, N. A. & Creelman, C. D. (1990). *Detection theory: A user's guide*. New York: Cambridge University Press.

Marcus, G. F. (2001). *The algebraic mind: integrating connectionism and cognitive science*. Cambridge, MA: The MIT Press.

Marx, M. H. & Cronan-Hillix, W. A. (1987). *Systems and theories in psychology* (4th ed.). New York: McGraw-Hill Book Company.

Maxwell, N. (2000). The mind-body problem and explanatory dualism. *Philosophy*, 75, 49–71.

McCauley, R. N. & Bechtel, W. (2001). Explanatory pluralism and heuristic identity theory. *Theory & Psychology*, 11, 736–760.

McFarland, D. (1999). *Animal behaviour: Psychology, ethology and evolution*. Harlow, UK: Longman.

McFee, G. (2000). *Free will*. Montreal & Kingston: McGill-Queen's University Press.

McGinn, C. (1982). *The character of mind*. Oxford: Oxford University Press.

McGinn, C. (1989). Can we solve the mind-body problem? *Mind*, 98, 349–366.

McGinn, C. (1991). *The problem of consciousness: Essays towards a resolution*. Oxford: Blackwell.

McLaughlin, B. P. (1992). The rise and fall of British emergentism. In A. Beckermann, H. Flohr & J. Kim (Eds.), *Emergence or reduction? Essays on the prospects of nonreductive physicalism*. Berlin: Walter de Gruyter

Mele, A. R. (2003). *Motivation and agency*. Oxford: Oxford University Press.

Meltzoff, A. N. (1996). The human infant as imitative generalist: A 20-year progress report of infant imitation with implications for comparative psychology. In C. M. Heyes & B. G. Galef (Eds.), *Social learning in animals: The roots of culture* (pp. 347–370). San Diego, CA: Academic Press.

Michell, J. (1990). *An introduction to the logic of psychological measurement*. Hillsdale, NJ: LEA.

Michell, J. (1999). *Measurement in psychology: Critical history of a methodological concept*. Cambridge: Cambridge University Press.

Mill, J. S. (1865). *System of logic*. London: Longmans, Green.

Millgram, E. (1997). *Practical induction*. Cambridge, MA: Harvard University Press.

Millikan, R. G. (1984). *Language, thought, and other biological categories.* Cambridge, MA: The MIT press.

Mitchell, R. W., Thompson, N. S. & Miles, H. L. (Eds.) (1997a). *Anthropomorphism, anecdotes, and animals.* New York: State University of New York Press.

Mitchell, R. W., Thompson, N. S. & Miles, H. L. (1997b). Taking anthropomorphism and anecdotes seriously. In R. W. Mitchell, N. S. Thompson & H. L. Miles (Eds.), *Anthropomorphism, anecdotes, and animals* (pp. 3–11). New York: State University of New York Press.

Morgan, C. L. (1977/1894). *An introduction to comparative psychology* (edited by Daniel N. Robinson). Washington DC: University Publications of America.

Morris, D. (1986). *Catwatching.* London: Jonathan Cape.

Morris, D. (1997). *Cat world: A feline encyclopedia.* New York: Penguin Reference.

Mowrer, O. H. (1947). On the dual nature of learning: A reinterpretation of "conditioning" and "problem-solving". *Harvard Educational Review,* 17, 102–150.

Nagel, E. (1961). *The structure of science: Problems in the logic of scientific explanation.* London: Routledge & Kegan Paul.

Nagel, T. (1974). What is like to be a bat? *The Philosophical Review,* 83, 435–450.

Neal, J. M. & Liebert, R. M. (1986). *Science and behavior: Introduction to methods of research* (3rd ed.). New Jersey: Prentice-Hall.

Nichols, S. (2004). Folk concepts and intuitions: From philosophy to cognitive science. *Trends in Cognitive Sciences,* 11, 514–518.

Nisbett, R. E. & Wilson, T. D. (1977). Telling more than we can know: Verbal reports on mental processes. *Psychological Review,* 84, 231–254.

O'Conner, T. (2002). Libertarian views: Dualist and agent-causal theories. In R. Kane (Ed.), *The Oxford handbook of free will.* Oxford: Oxford University Press.

Owens, D. (1989). Levels of explanation. *Mind,* 98, 59–79.

Pachella, R. G. (1974). The interpretation of reaction time in information-processing research. In B. H. Kantowitz (Ed.), *Human information processing: Tutorials in performance and cognition.* New York: LEA.

Palmer, S. E. & Kimchi, R. (1986). The information processing approach to cognition. In T. J. Knapp & L. C. Robertson (Eds.), *Approaches to cognition: Contrasts and controversies.* New Jersey: LEA.

Piggins, D. & Phillips, C. J. C. (1998). Awareness in domesticated animals – concepts and definitions. *Applied Animal Behaviour Science,* 57, 181–200.

Pinchin, C. (1990). *Issues in philosophy.* Savage, MD: Barnes & Noble.

Pitt, J. (Ed.) (1988). *Theories of explanation.* New York: Oxford University Press.

Poletiek, F. H. (2001). *Hypothesis-testing behaviour.* Sussex: Psychology Press.

Polger, T. W. (2004). *Natural minds.* Cambridge, MA: The MIT Press.

Popper, K. R. (1972/1934). *The logic of scientific discovery.* New York: Wiley.

Povinelli, D. J. (1998). Animal self-awareness: A debate – can animals empathize? Maybe not. *Scientific American Presents: Exploring Intelligence,* 9, 72–75.

Povinelli, D. J. & Giambrone, S. (1999). Inferring other minds: Failure of the argument by analogy. *Philosophical Topics,* 27, 167–201.

Psillos, S. (2002). *Causation and explanation.* Chesham, UK: Acumen.

Putnam, H. (1967). Psychological predicates. In W. Capitan & D. Merrill (Eds.), *Art, mind, and religion.* Pittsburgh, PA: University of Pittsburgh Press.

Putnam, H. (1975). *Mind, language and reality: Philosophical papers.* New York: Cambridge University Press.

Pylyshyn, Z. W. (1984). *Computation and cognition.* Cambridge, MA: The MIT press.
Radner, D. & Radner, M. (1989). *Animal consciousness.* New York: Prometheus Books.
Rakover, S. S. (1975). Tolerance of pain as a measure of fear. *Learning and Motivation,* 6, 43–61.
Rakover, S. S. (1979). Fish (tilapia aurea), as rats, learn shuttle better than lever-bumping (press) avoidance tasks: A suggestion for functionally similar universal reactions to conditioned fear-arousing stimulus. *American Journal of Psychology,* 92, 489–495.
Rakover, S. S. (1980). Role of intertrial interval following escape or avoidance response in bar-press avoidance. *Learning and Motivation,* 11, 220–237.
Rakover, S. S. (1983a). Hypothesizing from introspections: A model for the role of mental entities in psychological explanation. *Journal for the Theory of Social Behavior,* 13, 211–230.
Rakover, S. S. (1983b). In defense of memory viewed as stored mental representation. *Behaviorism,* 11, 53–62.
Rakover, S. S. (1986). Breaking the myth that behaviorism is a trivial science. *New Ideas in Psychology,* 4, 305–310.
Rakover, S. S. (1990). *Metapsychology: Missing links in behavior, mind and science.* New York: Paragon/Solomon.
Rakover, S. S. (1992). Outflanking the mind-body problem: Scientific progress in the history of psychology. *Journal for the Theory of Social Behavior,* 22, 145–173.
Rakover, S. S. (1993). Empirical criteria for task susceptibility to introspective awareness and awareness effects. *Philosophical Psychology,* 6, 451–467.
Rakover, S. S. (1996). The place of consciousness in the information processing approach: The mental-pool thought experiment. *Behavioral and Brain Sciences,* 19, 535–536.
Rakover, S. S. (1997). Can psychology provide a coherent account of human behavior? A proposed multiexplanation-model theory. *Behavior and Philosophy,* 25, 43–76.
Rakover, S. S. (1999). The computer that simulated John Searle in the Chinese Room. *New Ideas in Psychology,* 17, 55–66.
Rakover, S. S. (2002). Scientific rules of the game and the mind/body: A critique based on the theory of measurement. *Journal of Consciousness Studies,* 9, 52–58.
Rakover, S. S. (2003). Experimental psychology and Duhem's problem. *Journal for the Theory of Social Behaviour,* 33, 45–66.
Rakover, S. S. & Cahlon, B. (2001). *Face recognition: Cognitive and computational processes.* Amsterdam/Philadelphia: John Benjamins.
Reber, A. S. (1993). *Implicit learning and tacit knowledge: An essay on the cognitive unconsciousness.* New York: Oxford University Press.
Robinson, H. (2003). Dualism. In S. P. Stich & T. A. Warfield (Eds.), *The Blackwell guide to philosophy of mind.* Malden, MA: Blackwell.
Romanes, G. J. (1977/1883). *Animal intelligence* (edited by Daniel N. Robinson) Washington, DC: University Publications of America.
Rosenberg, A. (1988). *Philosophy of social science.* Boulder, CO: Westview Press.
Rosenberg, A. (2000). *Philosophy of science: A contemporary introduction.* London: Routledge.
Rosenthal, D. M. (1991). The independence of consciousness and sensory quality. In E. Villanueva (Ed.), *Consciousness.* Atascadero, CA: Ridgeview.
Ruben, D. (Ed.) (1993). *Explanation.* Oxford: Oxford University Press.
Salmon, W. C. (1967). *The foundation of scientific inference.* Pittsburgh: University of Pittsburgh Press.

Salmon, W. C. (1971). Statistical explanation. In W. Salmon (Ed.), *Statistical explanation and statistical relevance* (pp. 29–87). Pittsburgh: University of Pittsburgh Press.

Salmon, W. C. (1984). *Scientific explanation and the causal structure of the world.* Princeton: Princeton University Press.

Salmon, W. C. (1989) *Four decades of scientific explanation.* Minneapolis: University of Minneapolis Press.

Salmon, W. C. (1992). Scientific explanation. In M. H. Salmon, J. Earman, C. Glymour, J. G. Lennox, P. Machamer, J. E. McGuire, J. D. Norton, W. C. Salmon & K. F. Schaffner. *Introduction to the philosophy of science.* New Jersey: Prentice Hall.

Salmon, W. C. (1998). *Causality and explanation.* New York: Oxford University Press.

Salmon, W. C. (2001a). Explanation and confirmation: A Bayesian critique of inference to the best explanation. In G. Hon & S.S. Rakover (Eds.), *Explanation: Theoretical approaches and applications* (pp. 61–91). Dordrecht: Kluwer Academic Publishers.

Salmon, W. C. (2001b). Reflections of a bashful Bayesian: A reply to Lipton. In G. Hon & S.S. Rakover (Eds.,) *Explanation: Theoretical approaches and applications* (pp. 121–136). Dordrecht: Kluwer Academic Publishers.

Saygin, A. P., Cicekli, I. & Akman, V. (2000). Turing test: 50 years later. *Minds and Machines,* 10, 463–518.

Sayre-McCord, G. (1989). Functional explanations and reasons as causes. In J. E. Tomberlin (Ed.), *Philosophical perspectives. 3: Philosophy of mind and action theory.* Atascalero, CA: Ridgeview.

Schacter, D. L. (1989). On the relation between memory and consciousness: Dissociable interactions and conscious experience. In H. Roediger & F. Craik (Eds.), *Varieties of memory and consciousness: Essays in honor of Endel Tulving.* Hillsdale, NJ: LEA.

Schaffner, K. F. (1967). Approaches to reduction. *Philosophy of Science,* 34, 137–147.

Schaffner, K. F. (1993). *Discovery and explanation in biology and medicine.* Chicago: University of Chicago Press.

Searle, J. R. (1980). Minds, brains, and programs. *The Behavioral and Brain Sciences,* 3, 417–457.

Searle, J.R. (1992). *The rediscovery of the mind.* Cambridge, MA: The MIT Press.

Shear, J. (Ed.) (1997). *Explaining consciousness – the 'hard problem'.* Cambridge, MA: The MIT Press.

Shettleworth, S. J. (1998). *Cognition, evolution, and behavior.* Oxford: Oxford University Press.

Shiffrin, R. M. & Schneider, W. (1977). Controlled and automatic human information processing: II. Perceptual learning, automatic attending and a general hypothesis. *Psychological Review,* 84, 127–190.

Siewert, C. P. (1998). *The significance of consciousness.* Princeton, NJ: Princeton University Press.

Silberstein, M. (2002). Reduction, emergence and explanation. In P. Machamer & M. Silberstein (Eds.), *The Blackwell guide to the philosophy of science.* Malden, MA: Blackwell.

Silverman, P. S. (1997a). A pragmatic approach to the inference of animal mind. In R. W. Mitchell, N. S. Thompson & H. L. Miles (Eds.), *Anthropomorphism, anecdotes, and animals* (pp. 170–185) New York: State University of New York Press.

Simon, H. (1969). *The sciences of the artificial* (2nd ed.). Cambridge, MA: The MIT Press.

Simon, H. (1973). The organization of complex systems. In H. H. Pattee (Ed.), *Hierarch theory: The challenge of complex systems* (pp. 1–27). New York: Braziller.

Sklar, L. (1967). Types of inter-theoretic reductions. *British Journal for the Philosophy of Science,* 18, 109–124.

Smilansky, S. (2000). *Free will and illusion*. Oxford: Clarendon Press.
Smith, J. D., Shields, W. E. & Washburn, D. A. (2003). The comparative psychology of uncertainty monitoring and metacognition. *Behavioral and Brain Sciences*, 26, 317–373.
Smolensky, P. (1988). On the proper treatment of connectionism. *Behavioral and Brain Sciences*, 11, 1–74.
Stich, S. P. (1983). *From folk psychology to cognitive science*. Cambridge, MA: The MIT Press.
Stich, S. P. & Nichols, S. (2003). Folk psychology. In S. P. Stich & T. A. Warfield (Eds.), *The Blackwell guide to philosophy of mind* (pp. 235–255). Malden, MA: Blackwell.
Stich, S. P. & Ravenscroft, I. (1994). What is folk psychology? *Cognition*, 50, 447–468.
Stout, R. (1996). *Things that happen because they should: A teleological approach to action*. Oxford: Oxford University Press.
Suarez, M. (2004). An inferential conception of scientific representation. *Philosophy of Science*, 71, 767–7792.
Suppe, F. (1977). Introduction. In F. Suppe (Ed.), *The structure of scientific theories* (pp. 3–241). Urbana, IL: University of Illinois Press.
Suppe, F. (1984). Beyond Skinner and Kuhn. *New Ideas in Psychology*, 2, 89–104.
Swartz, N. (1985). *The concept of physical law*. Cambridge: Cambridge University Press.
Tabor, R. (1997). *Roger Tabor's cat behaviour: The complete feline problem solver*. Newton Abbot: David & Charles.
Tavolga, W. N. (1969). *Principles of animal behavior*. New York: Harper & Row.
Taylor, D. (1986). *You and your cat, the complete owner's guide to cats: Their care, health and behavior*. London: Dorling Kindersley.
Thagard, P. (1996). *Mind: Introduction to cognitive science*. Cambridge, MA: The MIT Press.
Thomas, R. K. (1998). Lloyd Morgan's canon. In G. Greenberg & M. M. Haraway (Eds.), *Comparative psychology: A handbook* (pp.156–163). New York: Garland.
Tinbergen. N. (1951/1969). *The study of instinct*. Oxford: Oxford University Press.
Tinbergen, N. (1963). On the aims and methods of ethology. *Zeitschrift fur Tierpsychologie*, 20, 410–433.
Tinbergen, N. & Editors of Time-Life Books (1965). *Animal behavior*. New York: Time-Life Books.
Trout, J. D. (2002). Scientific explanation and the sense of understanding. *Philosophy of Science*, 69, 212–233.
Tye, M. (1996). *Ten problems of consciousness: A representational theory of the phenomenal mind*. Cambridge, MA: The MIT Press.
Ulrich, R. (1966). Pain as a cause of aggression. *American Zoology*, 6, 643–662.
Uttal, W. R. (2001). *The new phrenology: The limits of localizing cognitive processes in the brain*. Cambridge, MA: The MIT Press.
van Fraassen, B. C. (1980). *The scientific image*. Oxford: Clarendon Press.
van Fraassen, B. C. (2000). The theory of tragedy and of science: Does nature have narrative structure? In D. Sfendoni-Metzou (Ed.), *Aristotle and contemporary science* (vol. 1). New York: Peter Lang.
van Fraassen, B. C. (2004). Scientific representation: Flouting the criteria. *Philosophy of Science*, 71, 794–804.
Van Gulick, R. (1995). What would count as explaining consciousness? In T. Metzinger (Ed.), *Conscious experience*. Schoningh: Imprint Academic.

Velmans, M. (1991). Is human information processing conscious? *Behavioral and Brain Sciences*, 14, 651–669.

Von Eckhardt, B. (1993). *What is cognitive science?* Cambridge, MA: The MIT Press.

Von Wright, G.H. (1971). *Explanation and understanding.* London: Routledge & Kegan Paul.

Weinert, F. (1995). Laws of nature – laws of science. In F. Weinert (Ed.), *Laws of nature: Essays on the philosophical, scientific and historical dimensions*. Berlin; New York: de Gruyer.

Wimsatt, W. C. (1972). Complexity and organization. In K. Schaffner & R. S. Cohen (Eds.), *PSA* 2, pp. 67–86. Dordrecht: D. Reidel.

Woodward, J. (2000). Explanation and invariance in the special sciences. *British Journal for the Philosophy of Science*, 51, 197–254.

Woodward, J. (2002). Explanation. In P. Machamer & M. Silberstein (Eds.), *The Blackwell guide to the philosophy of science.* Oxford: Blackwell.

Woodward, J. (2003). *Making things happen: A theory of causal explanation.* Oxford: Oxford University Press.

Woody, A. I. (2004) More telltale signs: What attention to representation reveals about scientific explanation. *Philosophy of Science*, 71, 780–793.

Subject index

Name index

Advances in Consciousness Research

A complete list of titles in this series can be found on the publishers' website, *www.benjamins.com*

71 **KROIS, John Michael, Mats ROSENGREN, Angela STEIDELE and Dirk WESTERKAMP (eds.):** Embodiment in Cognition and Culture. 2007. xxii, 304 pp.

70 **RAKOVER, Sam S.:** To Understand a Cat. Methodology and philosophy. 2007. xvii, 253 pp.

69 **KUCZYNSKI, John-Michael:** Conceptual Atomism and the Computational Theory of Mind. A defense of content-internalism and semantic externalism. 2007. x, 524 pp.

68 **BRÅTEN, Stein (ed.):** On Being Moved. From mirror neurons to empathy. 2007. x, 333 pp.

67 **ALBERTAZZI, Liliana (ed.):** Visual Thought. The depictive space of perception. 2006. xii, 380 pp.

66 **VECCHI, Tomaso and Gabriella BOTTINI (eds.):** Imagery and Spatial Cognition. Methods, models and cognitive assessment. 2006. xiv, 436 pp.

65 **SHAUMYAN, Sebastian:** Signs, Mind, and Reality. A theory of language as the folk model of the world. 2006. xxvii, 315 pp.

64 **HURLBURT, Russell T. and Christopher L. HEAVEY:** Exploring Inner Experience. The descriptive experience sampling method. 2006. xii, 276 pp.

63 **BARTSCH, Renate:** Memory and Understanding. Concept formation in Proust's *A la recherche du temps perdu*. 2005. x, 160 pp.

62 **DE PREESTER, Helena and Veroniek KNOCKAERT (eds.):** Body Image and Body Schema. Interdisciplinary perspectives on the body. 2005. x, 346 pp.

61 **ELLIS, Ralph D.:** Curious Emotions. Roots of consciousness and personality in motivated action. 2005. viii, 240 pp.

60 **DIETRICH, Eric and Valerie Gray HARDCASTLE:** Sisyphus's Boulder. Consciousness and the limits of the knowable. 2005. xii, 136 pp.

59 **ZAHAVI, Dan, Thor GRÜNBAUM and Josef PARNAS (eds.):** The Structure and Development of Self-Consciousness. Interdisciplinary perspectives. 2004. xiv, 162 pp.

58 **GLOBUS, Gordon G., Karl H. PRIBRAM and Giuseppe VITIELLO (eds.):** Brain and Being. At the boundary between science, philosophy, language and arts. 2004. xii, 350 pp.

57 **WILDGEN, Wolfgang:** The Evolution of Human Language. Scenarios, principles, and cultural dynamics. 2004. xii, 240 pp.

56 **GENNARO, Rocco J. (ed.):** Higher-Order Theories of Consciousness. An Anthology. 2004. xii, 371 pp.

55 **PERUZZI, Alberto (ed.):** Mind and Causality. 2004. xiv, 235 pp.

54 **BEAUREGARD, Mario (ed.):** Consciousness, Emotional Self-Regulation and the Brain. 2004. xii, 294 pp.

53 **HATWELL, Yvette, Arlette STRERI and Edouard GENTAZ (eds.):** Touching for Knowing. Cognitive psychology of haptic manual perception. 2003. x, 322 pp.

52 **NORTHOFF, Georg:** Philosophy of the Brain. The brain problem. 2004. x, 433 pp.

51 **DROEGE, Paula:** Caging the Beast. A theory of sensory consciousness. 2003. x, 183 pp.

50 **GLOBUS, Gordon G.:** Quantum Closures and Disclosures. Thinking-together postphenomenology and quantum brain dynamics. 2003. xxii, 200 pp.

49 **OSAKA, Naoyuki (ed.):** Neural Basis of Consciousness. 2003. viii, 227 pp.

48 **JIMÉNEZ, Luis (ed.):** Attention and Implicit Learning. 2003. x, 385 pp.

47 **COOK, Norman D.:** Tone of Voice and Mind. The connections between intonation, emotion, cognition and consciousness. 2002. x, 293 pp.

46 **MATEAS, Michael and Phoebe SENGERS (eds.):** Narrative Intelligence. 2003. viii, 342 pp.

45 **DOKIC, Jérôme and Joëlle PROUST (eds.):** Simulation and Knowledge of Action. 2002. xxii, 271 pp.

44 **MOORE, Simon C. and Mike OAKSFORD (eds.):** Emotional Cognition. From brain to behaviour. 2002. vi, 350 pp.

43 **DEPRAZ, Nathalie, Francisco J. VARELA and Pierre VERMERSCH:** On Becoming Aware. A pragmatics of experiencing. 2003. viii, 283 pp.

42 **STAMENOV, Maxim I. and Vittorio GALLESE (eds.):** Mirror Neurons and the Evolution of Brain and Language. 2002. viii, 392 pp.

41 **ALBERTAZZI, Liliana (ed.):** Unfolding Perceptual Continua. 2002. vi, 296 pp.

40 **MANDLER, George:** Consciousness Recovered. Psychological functions and origins of conscious thought. 2002. xii, 142 pp.

39 **BARTSCH, Renate:** Consciousness Emerging. The dynamics of perception, imagination, action, memory, thought, and language. 2002. x, 258 pp.